HANDBOOK FOR ROOFTOP SOLAR DEVELOPMENT IN ASIA

ADB

ASIAN DEVELOPMENT BANK

ISBN 978-92-9254-847-6 (Print), 978-92-9254-848-3 (e-ISBN)
Publication Stock No. TIM146676

Cataloging-In-Publication Data

Asian Development Bank.
        Handbook for rooftop solar development in Asia.
Mandaluyong City, Philippines: Asian Development Bank, 2014.

1. Solar photovoltaic system.   2. Energy.   3. Clean energy.   I. Asian Development Bank.

Note:
In this publication, "$" refers to US dollars.

6 ADB Avenue, Mandaluyong City
1550 Metro Manila, Philippines
Tel +63 2 632 4444
Fax +63 2 636 2444
www.adb.org

For orders, please contact:
Public Information Center
Fax +63 2 636 2584
adbpub@adb.org

Contents

Contents

Boxes, Figures, and Tables

Preface

This *Handbook for Rooftop Solar Development in Asia* was written to support the Asian Development Bank's (ADB) Asia Solar Energy Initiative (ASEI), which aims to create a virtuous cycle of solar energy investments in the region, so that developing Asian countries may optimally benefit from the clean and inexhaustible energy provided by the sun.

The ASEI uses an integrated, multipronged approach that features three interlinked components: (1) knowledge management, (2) project development, and (3) innovative finance solutions. This publication furthers the first prong of disseminating knowledge management solutions, to facilitate solar technology transfer, deliver quality inputs to policy and decision makers in solar power development, and facilitate innovation in financing mechanisms.

In addition to catalyzing large, utility-scale solar projects, the Asia and Pacific region will also benefit from further exploring the possibilities of rooftop solar photovoltaic (PV) technology. It has particular utility in being suited for decentralized solar power generation for remote and rural communities, although this publication also shows that larger-scale urban systems are practical, economical, and make good use of unused rooftop space.

As such, this handbook demystifies the process of implementing a rooftop solar PV project through a step-by-step guide to development. It covers the initial stages of how to conduct a prefeasibility assessment, how to finance a system, and how to ensure proper operations and maintenance. By being aware of all the steps in the process, and with guidance on handling the various aspects, this handbook hopes to help streamline the implementation process and therefore make rooftop solar PV much more accessible.

As a point of reference, this handbook provides examples from ADB's experience in implementing its own rooftop solar PV system, which ADB completed in 2012. Though this handbook was written to be accessible to all audiences, experienced developers may find ADB's experience helpful, as we have included tips for bidding and procurement, and in choosing an adequate financing mechanism.

With the cost of solar PV systems decreasing, alongside rising fossil fuel and electricity prices, implementing a rooftop solar PV system is becoming an attractive option, even in heavily urbanized areas. ADB has predicted increased energy demand in Asia's future, and rooftop solar PV is one option, among many alternative energy solutions, that can meet that demand in a sustainable manner.

Anthony Jude
Senior Advisor concurrently Practice Leader (Energy)
Asian Development Bank

Acknowledgments

This publication is an output of the Clean Energy Program of the Asian Development Bank (ADB). It is a multipronged initiative that seeks to increase regional energy efficiency in the energy, transport, and urban sectors; adopt renewable energy sources; and improve access to energy. Providing access to clean energy helps achieve ADB's mission of a region free of poverty.

From ADB, Aiming Zhou, senior energy specialist, supervised the preparation of the handbook. Carina Paton prepared the handbook, with assistance from Chatiya Nantham, Alejandro Ana, Oscar R. Roman, Anil Terway, and Mark Lister who provided research and technical support. Nelly Sangrujiveth drafted the policy and regulations section and provided editorial assistance. Staff support was provided by Maria Dona Aliboso, Patricia Calcetas, and Maria Angelica Rongavilla. Andres Kawagi Fernan and Charito Isidro provided valuable help at various stages of the production and printing process.

Roberto R. Martin of Propmech provided technical information pertaining to the ADB installation. Ericson B. Maquinto, Melvin B. Atienza, and Graziel Joy G. Evangelio of TBS Industrial Services also shared their technical expertise.

ADB extends its gratitude toward the nine members of the Institutional Procurement Committee, who managed the procurement for the ADB rooftop solar project and provided insight for this publication. These members include Akmal Siddiq (chairperson), Paulino Santiago (vice-chair and secretary), Cesar Valbuena (member), Oliver Leonard (member), Aiming Zhou (member and technical representative), Paul Hattle (member and technical representative), Gicheon Balk (independent member), Donald Kidd (legal advisor), and Chatiya Nantham (member).

ADB also extends its gratitude toward the experts who provided additional insight into policies and regulations for implementing rooftop solar PV in the Philippines: Director Mario C. Marasigan and Chief Advisor Hendrik Muller of the Philippine Department of Energy's Renewable Energy Management Bureau.

Abbreviations

AC	alternating current
ACEGE	AC electricity generation effectiveness
ADB	Asian Development Bank
BIPV	building-integrated photovoltaics
CDM	Clean Development Mechanism
CdTe	cadmium-telluride
CER	carbon emission reduction credit
CIGS	copper-indium-gallium-selenide
CPV	concentrating photovoltaics
DAS	data acquisition system
DC	direct current
DHI	diffuse horizontal irradiance
DMC	developing member country
DNI	direct normal irradiance
DSSC	dye-sensitized solar cell
FIT	feed-in tariff
GHI	global horizontal irradiance
GSMC	Good Social Management Certificate
HVAC	heating, ventilation, and air conditioning
IEA	International Energy Agency
IEC	International Electrotechnical Commission
LEED	Leadership in Energy and Environmental Design
NREL	National Renewable Energy Laboratory
OPV	organic photovoltaics
PPA	power purchase agreement
PR	performance ratio
PV	photovoltaic
REC	renewable energy certificate
RPS	renewable portfolio standard
SLA	solar leasing arrangement
STC	standard test conditions
UL	Underwriters Laboratories
UNFCCC	United Nations Framework Convention on Climate Change
US DOE	United States Department of Energy

Weights and Measures

A	ampere
°C	degree Celsius
GWh	gigawatt-hour
kg	kilogram
kW	kilowatt
kWh	kilowatt-hour
kWp	kilowatt-peak
m²	square meter
MJ	megajoules
MW	megawatt
MWh	megawatt-hour
V	volt
W	watt
W/m²	watts per square meter
Wp	peak watt

Introduction

Purpose of Handbook

The Handbook for Rooftop Solar Development in Asia intends to serve as a reference for potential project developers (including property owners and financiers) and contractors who are considering rooftop solar photovoltaic (PV) systems. It answers questions that people encounter along the way, including issues dealing with design considerations before installing a PV system, the permits required, available incentives and financing options, procurement, installation considerations including safety issues, and operation and maintenance of the system.

This handbook also covers the basics of solar PV systems. It includes background information on solar resources, and the types of PV technologies and mounting systems most commonly installed on commercial buildings.

To better illustrate how to implement the steps for implementing a rooftop solar PV system, this handbook provides examples from the Asian Development Bank's (ADB) own experience with its Headquarters Rooftop Solar Power Project in Manila. It is representative of how to implement a rooftop solar PV system cost-effectively, while fulfilling ADB's goals for supporting efforts to provide Asia and the Pacific with affordable and sustainable sources of energy. Providing detailed knowledge of this project seeks to inspire others to also transform their rooftops.

Rooftop Solar: A High-Benefit Power Source

Present-day solar PV technology, a low-carbon energy solution, is well suited for much of Asia and the Pacific. With large areas of the region endowed with bountiful solar radiation, many countries in the region have the ideal conditions for utilizing solar energy.

Most solar PV systems tend to be one of two types. The first type are utility-scale installations with a capacity usually above 1 megawatt (MW). They require large, open land areas with few shadows. The second type is distributed generation, which may be ground-mounted or installed on rooftops. They generate power during the day, while feeding surplus power back into the power grid. Residences can be sufficiently supplied with small systems of usually up to 20 kilowatts (kW), while larger public, commercial, and industrial buildings may have systems with a capacity as large as 1 MW.

Although much smaller in capacity than power plant-type installations, the rooftop solar system has many benefits in helping us change how we produce energy and make our world a better place to live. The benefits are summarized in Table 1.

Table 1: Benefits of Rooftop Solar Power

Construction	
Site access	Photovoltaic (PV) systems are at the point of consumption, thus do not require additional investment for access during construction or for operation and maintenance.
Modularity	They can be designed for easy expansion if power demand increases.
Operation and Maintenance	
Primary energy supply	Solar energy is freely available, and the PV system does not entail environmental costs for conversion to electricity.
Maintenance	PV systems require little maintenance.
Peak generation	These systems offset the need for grid electricity generation to meet expensive peak demand during the day.
Mature technology	PV systems nowadays are based on proven technology that has operated for over 25 years.
Impact	
Investments	Rooftop PV system costs help offset part of the investment needed for new power generation, transmission, and distribution in the power grid.
Cost	Fuel savings from PV systems typically offset their relatively high initial cost.
Environment	PV systems create no pollution or waste products while operating, and production impacts are far outweighed by environmental benefits.

Source: ADB.

Case Study: How ADB Transformed Its Rooftop

To demonstrate the types of choices developers can make in implementing a rooftop solar PV system, this handbook references ADB's experience in implementing its Headquarters Rooftop Solar Power Project. As such, this section provides some preliminary background information on that project. More of its details are revealed throughout this handbook where relevant.

As an institution that promotes sustainable development within its developing member countries (DMCs), ADB sought to lead by example by transforming its 23-year-old headquarters building in Manila into a showcase for sustainability. The idea was that if ADB could cost-effectively transform a building of this age to meet present-day Leadership in Energy and Environmental Design (LEED) certification standards,[1] that the transformation would inspire others to follow suit.

Thus, beginning in 2007, ADB embarked on a number of retrofits, which started with making energy efficiency improvements. Efficient ventilation and air conditioning, lighting, and other technologies have since enabled ADB to reduce energy consumption by 4%. A newly-constructed car park building also sought to be "green," by utilizing a solar PV system and batteries for lighting.

---

[1] LEED certification refers to a green building ratings system, as developed by the US Green Building Council. Upon installation of the ADB Rooftop Solar Power Project, ADB received a LEED Gold rating from the U.S. Green Building Council.

Coinciding with these efficiency improvements were plans to install a large distributed rooftop solar PV system in ADB's main headquarters. The decision was based on a number of reasons.

First, ADB understands the benefits of solar energy. ADB has been actively promoting this technology in its DMCs under the Asia Solar Energy Initiative, which was launched in 2010. ADB wanted to lead by example and produce solar energy at ADB headquarters.

Second, ADB needed to showcase how commercial buildings in Asia can reduce their carbon footprint and diversify energy supply with renewable energy. The project preparation, approval, financing, and implementation had to be easily replicable and scalable for enabling wider use of solar energy.

Finally, ADB took the opportunity to start a market for solar PV systems. As more commercial buildings implement similar rooftop projects, costs will reduce and deployment of the technology will accelerate.

Since the project's successful completion in 2012, ADB has been encouraging like-minded enterprises in Manila and elsewhere in Asia to install similar facilities on their rooftops and other available spaces. With energy demand projected to almost double in the region by 2030, ADB envisions that these pioneering projects will enable the power industry to meet new demand with reduced carbon impact. Reducing the region's reliance on fossil fuel is a necessary step toward enhancing future energy security and mitigating climate change.

Other enterprises have since followed suit, as ADB's rooftop solar project has proven to be a cost-effective and reliable power source. The system has a capacity of 571 kW, and generates 50,000 kWh of electricity per month under average weather conditions. This is enough to power 245 Metro Manila households using an average of 2,500 kilowatt-hours (kWh) per year. The solar-generated electricity supplements ADB's purchase of geothermal-generated electricity from AdventEnergy, which supplies an average of 1.5 gigawatt-hours (GWh) of electricity per month.

Thus, the ADB Headquarters building is powered by 100% renewable energy. Making that transition has enabled ADB to cut its annual carbon footprint by 50% and reduced its emissions by more than 9,500 tons of carbon dioxide equivalent.

Handbook Contents

This handbook breaks down the development of rooftop solar PV systems into five chapters: (1) project preparation, (2) system design, (3) procurement, (4) implementation, and (5) operation and maintenance. These chapters correspond to the five different stages of project development.

Figure 1 depicts a flowchart of the sequence of events in developing a rooftop solar project. As depicted in the flowchart, some of these stages will inherently overlap with each other. The subsequent sections will denote when these overlaps may happen, as relevant.

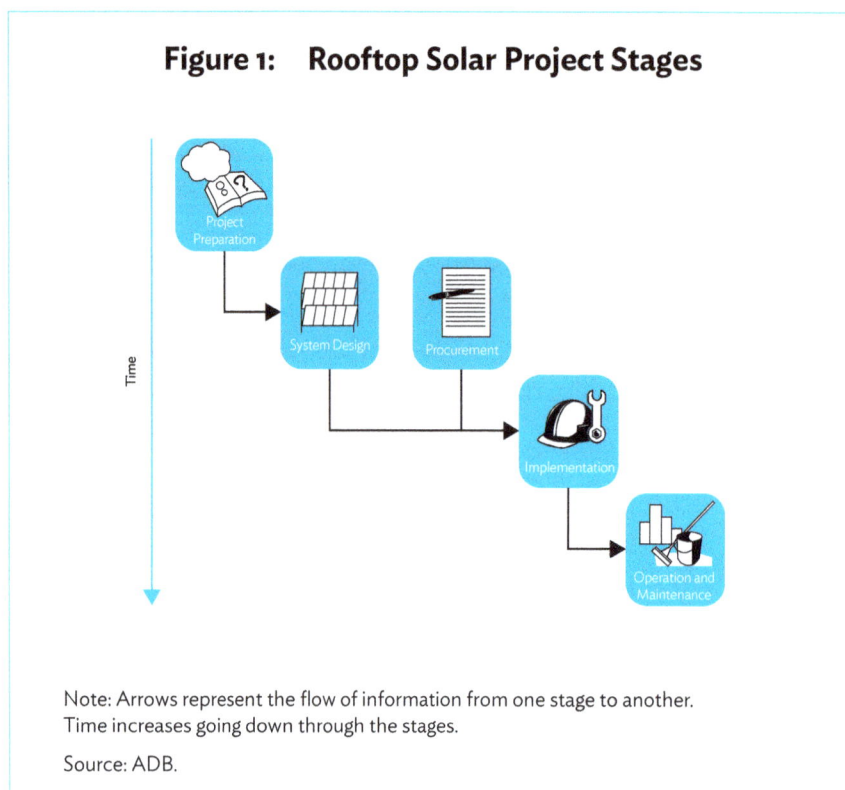

Figure 1: Rooftop Solar Project Stages

Note: Arrows represent the flow of information from one stage to another.
Time increases going down through the stages.

Source: ADB.

Chapter 1: Project Preparation

This chapter gives several key aspects developers should investigate to help decide if it is worth proceeding with a rooftop solar PV project. The investigation includes assessments of the site conditions, legal and regulatory frameworks, business models and financing options, and implementation arrangements.

Chapter 2: System Design

This chapter outlines many of the system design aspects, including how to assess the site and resources, select suitable components for the system, configure the array, estimate system performance, and conduct due diligence.

Chapter 3: Procurement

This chapter describes a method to procure experts' services to design, install, and maintain the rooftop solar system: the one-stage, two-envelope procedure used for the rooftop system of ADB.

Chapter 4: Implementation

This chapter outlines the steps that should be taken to implement the project, including acquiring necessary permits and clearances, installing the various components, and the final testing and commissioning of the solar power system.

Chapter 5: Operation and Maintenance

This chapter contains activities to maintain and monitor performance of a solar PV system.

Chapter 1: Project Preparation

The first step in the project development process is to assess whether it is even worthwhile to proceed. Undertaking that assessment, which is also known as conducting a prefeasibility study, can evaluate a number of factors depending on the individual needs of the rooftop owner.

As seen with the ADB Headquarters Rooftop Solar Power Project, many of those factors may not be relevant to everyone (see Box 1). However, common factors that most people weigh heavily are technical feasibility and cost.

Measuring those two factors will require an evaluation of the following:

(i)   the site of the proposed project,
(ii) legal and regulatory frameworks,
(iii) permits and licensing,
(iv) financing options, and
(v)   implementation arrangements.

The sections below provide an overview of how to evaluate each of these five aspects, while keeping technical feasibility and cost in mind.

Box 1:   Project Preparation Activities for the ADB Headquarters Rooftop Solar Power Project

Because the Asian Development Bank (ADB) sought to share its experience with its developing member countries, its prefeasibility study evaluated more than technical and cost factors. The prefeasibility study investigated the following aspects, some of which are further detailed in Boxes 2–6:

(i)   **Rationale for the project.** How does it align with institutional strategies and goals?
(ii)   **Knowledge product to be disseminated.** What information will be produced and shared with the public about the project?
(iii) **Project specifications.** How much energy is consumed in the headquarters building? What are the basic specifications of the selected roof area? What solar photovoltaic system capacity may be accommodated?
(iv)   **Proposed business model and procurement process.** Who will own and operate the equipment, and how will contractors be selected?
(v)   **Project cost and financing plan.** How much will the project cost, and what are the financing options?
(vi)   **Implementation arrangements and schedule.** Who will be involved in the project, and what is the proposed schedule?
(vii) **Risk assessment.** What are the technology and business risks of the project?
(viii) **Legal and regulatory requirements.** What are ADB's legal requirements? How does the renewable energy law pertain to this project?
(ix)   **Requirements for permits and clearances.** What permits and clearances are needed from the local and national government bodies, and what is the application process?

Source: ADB.

Conducting a prefeasibility study may require assembling a project team. Should that be the case, clearly defining its organizational structure and each team member's responsibilities becomes useful. For the ADB Headquarters Rooftop Solar Power Project, ADB set up a project team composed of representatives from all relevant divisions, and designated roles and responsibilities. ADB also set up a steering committee to guide the project team.

Another important aspect is establishing timelines for completing tasks. Even beyond the prefeasibility study, a roadmap of main milestones of the process and the expected dates of completion for each will help streamline implementation. Installation of a PV system can usually be carried out in a few months, although panel manufacturing and importing would require additional time.

1.1 Site Assessment

A well-conducted assessment of the roof requires developers to answer the following questions:

- *Is the roof suitable for installation of solar PV?*
- *Is the solar resource high enough?*
- *How much installed capacity could fit on the roof?*
- *How much energy could that system deliver?*

This section provides an overview of how to answer those questions.

1.1.1 Is the roof suitable for installation of solar PV?

A rooftop solar PV installation comprises of PV panels assembled in arrays, mounting frames to support the panels and secure them to the roof, wiring, inverters, and other components depending on the type of installation. The roof site must be able to accommodate all of these components, which requires examining the following aspects:

Accessibility. The roof must be accessible to carry out installation and maintenance. It must be possible to lift the solar system components onto the roof and for personnel to physically access the site to install and maintain the system.

Roof configuration. A roof plan can help quantify the roof area available for the PV power plant. The plan should indicate the location (including longitude and latitude), height, and slope of the roof itself, as well as any additional structures present on the roof. Identify any possible conflicts in usage of the roof, such as a helipad or communication antennae, and contact relevant bodies to ascertain if any special permission is required to use and/or alter usage of the roof space.

Roof materials and structure. For existing buildings, first find out when the roof would need replacement. If a roof is nearing the end of its life span, it is more cost-effective to install the rooftop PV system once the new roof is in place. It is also easier to integrate a system into the design of a new roof.

Next, engage a structural engineer to determine if the roof can be penetrated to secure the mounting frames. Often, flat roofs have a membrane that will lose its waterproofing properties if penetrated—in this case, the system may require a ballasted mounting system (with concrete weights).

The engineer should also be able to determine if the roof can bear the additional weight of a PV system. As a general rule of thumb, a crystalline PV system will place about 15–20 kilograms per square meter (kg/m^2) (3–4 pounds per square foot) of dead load on the roof (California Energy Commission 2001), but this varies depending on the panels and mounting system used, the spacing between panels, and the wind load.

Generally, a flat, concrete terrace roof will normally have the strength to accommodate the additional weight of the panels and supporting structures. Inclined roofs of storage sheds and residential buildings may be made of metal sheet, tiles, or similar materials, in which case it would be necessary to examine whether the trusses can support additional weight.

In the event that the roof is unable to support the load of a crystalline PV system, lighter thin-film modules could be an option.

Finally, if the roofing installation and manufacturing warranties are still valid, determine if installation of a rooftop solar system could void the warranty.

Shading. Nearby buildings or trees can shade a roof during certain periods of the day, which will lower the power output of the solar PV system. Although shading changes throughout a year, observing the roof at various times during a day gives a fair idea of the typical number of sunlight hours.

Aesthetics. Check that the solar PV modules would not negatively affect the aesthetics of the building. From street level, solar modules will be more visible on a sloped roof than on a flat roof. If they will be seen, find out if there are any local building restrictions preventing a visible rooftop solar PV installation. With growing support for the use of renewable energy, guidelines are being modified, where necessary, to allow rooftop solar installations.

Roof leasing. If planning to lease the roof space to the owner of the rooftop PV system, consult a legal advisor who would be able to confirm whether that type of arrangement is permissible.

Electrical load. Obtain the current and expected electrical load of the building or facility. Should the load be comparable to or less than the electricity generated through solar PV, plan a smaller system or plan to use the excess energy—either store the energy in batteries, send it to another building within the facility, or feed it into the grid.

These aspects can easily be incorporated into the design of a new building. However, the investigation will take more effort for solar installations on a preexisting roof. Box 2 describes ADB's site assessment, in the context of the site being on the roof of an older building.

1.1.2   Is the solar resource high enough?

A national or regional solar energy map showing global horizontal radiation will indicate approximately how much solar resource is available in any given location.[2] It depicts a measurement of the solar radiation that reaches the earth's surface. It can be expressed as either Global Horizontal Irradiance or Global Horizontal Irradiation. Global Horizontal Irradiation measures irradiance over a certain time period, such as one year. Box 3 instructs how to convert a measurement of Global Horizontal Irradiance to Global Horizontal Irradiation.

The worldwide annual average global horizontal irradiance (GHI)[3] is 170 watts per square meter (W/m²) (World Energy Council 2007).[4] However, many regions receive much more. Most of Southeast Asia receives an annual average GHI of 180–230 W/m² (equivalent to 1,600–2,000 kilowatt-hours per square meter per year [kWh/m²/year], see Box 3).[5] This is adequate for a solar PV system, considering that Germany, the global leader in solar PV installed capacity (REN21 2013), receives just over 1,300 kWh/m²/year even in the area with highest irradiation (the south of the country).[6] This rough estimation of GHI will be useful later when calculating predicted energy output from the proposed system.

---

[2]   National solar irradiance maps will typically provide higher resolution and accuracy than regional maps. They are often freely available from government agencies responsible for renewable energy development.

[3]   A full definition of GHI is given in the section on Chapter 2: System Design.

[4]   Instantaneous energy received from the sun is power, which varies between 0 and about 1,000 W/m² at midday. The average power is usually a quarter of maximum power.

[5]   3Tier Inc., www.3Tier.com

[6]   Based on a map by Solar GIS. *Global horizontal irradiation: Germany.* http://solargis.info/doc/_pics/freemaps/1000px/ghi/SolarGIS -Solar-map-Germany-en.png (accessed 4 February 2014).

Box 2:   Assessment of ADB Headquarters Rooftop Space

One of the main purposes of the Asian Development Bank (ADB) Rooftop Solar Power Project is to demonstrate that solar photovoltaic (PV) technology is ready for common use and thereby boost development of similar solar PV projects throughout the Asia-Pacific region.

To do so, in addition to making knowledge products like this handbook available, ADB wanted to make available public viewings of the rooftop solar power project. Thus, roof access was an important parameter.

Choosing to locate the project on the roof of the Facilities Block building of ADB Headquarters allowed accessibility, since visitors could access the rooftop via an internal staircase. The roof of the Facilities Block also had a larger installable area than the other roofs in ADB Headquarters, as well as fewer physical obstructions.

Expectedly, the roof of the Facilities Block had not been designed with a solar system in mind. The biggest obstacle was that it contains a waterproof membrane, which means the mounting method for the solar PV panels could not be intrusive.

However, the roof also has positive aspects for installing a rooftop PV system even though it was not designed with one in mind. For instance, it sits around relatively few tall buildings that would cast long shadows, so exposure to daylight hours would be long. Also, aesthetics were not an issue since the roof area is flat and solar PV panels would not be visible from the street level.

Source: ADB.

Box 3:   Global Horizontal Irradiation Conversion Guide

Global horizontal irradiation is measured in a variety of units in solar irradiation maps. Some common conversion factors are as follows:

_____	MJ/m²/day	÷	3.6x	=	_____	kWh/m²/day
_____	W/m²	×	0.024	=	_____	kWh/m²/day
_____	kWh/m²/day	×	365	=	_____	kWh/m²/year

where kWh = kilowatt-hour, m² = square meter, MJ = megajoule, W = watt.

Some maps display global horizontal irradiation "at latitude tilt," meaning that the pyranometer is tilted at the angle of the latitude—this will give a higher number than if it were horizontal. This effect is minimal at lower latitudes, and so latitude tilt maps are more common for areas further from the equator. The global irradiance at latitude tilt $(GI_\emptyset)$ for a site at latitude Ø can be converted to global horizontal irradiance (GHI) as follows:

$$GHI = \frac{GI_\emptyset}{\cos \emptyset}$$

Source: ADB.

1.1.3 How much installed capacity could fit on the roof?

An estimate of the potential installed capacity of the rooftop solar PV system, C_R, in kilowatt-peak (kWp), may be made using the following equation:

$$C_R = \left(\frac{C_M}{1{,}000}\right) \times \left(\frac{RCR \times A_R}{A_M}\right)$$

where:

- A_R is the total roof area available for installation of solar modules in m²,
- C_M is the individual module rated capacity in Wp,
- A_M is the area of one module in m², and
- RCR is the roof cover ratio, which is the fraction of roof area that the modules will cover.[7]

The roof area (A_R) should be the total area minus the footprint of any obstacles on the roof, such as helipads, water tanks, utility rooms, communication towers, and air conditioning systems.

For module rated capacity (C_M), we suggest obtaining up-to-date module specification sheets, since solar technology is continuously evolving. These module specification sheets are available for commercial solar modules from manufacturers and distributors.

A typical value for the roof cover ratio (RCR) is 0.85, which would allow for 15% of the roof to be free for spacing between modules and away from obstructions.

1.1.4 How much energy could the system deliver?

Energy yield (E) in kilowatt-hour can be estimated using the following equation:

$$E = C_R \times GHI_a \times D$$

In this equation:

- C_R is the potential installed capacity of the solar PV rooftop system.
- GHI_a is the global horizontal irradiation over a 1-year period in kWh/m². Values can be obtained from a national or regional solar map.[8]
- D is the derate factor for converting direct current (DC) to alternating current (AC). The derate factor generally ranges between 0.6 and 0.8 (Honda et al. 2012). A typical derate factor of 0.75 can be used here.

To further illustrate how to calculate energy yield, Box 4 provides an example using the ADB rooftop solar power project's estimations.

---

[7] Equation derived from ADB calculations.
[8] Technically, this should be global irradiance at the tilt angle of the panels; the GHI is sufficient for a quick estimate but at low latitudes.

Box 4:   Estimated Capacity of and Energy Delivered by the ADB Rooftop PV Project

The Asian Development Bank (ADB) examined the Facilities Block roof plans and found the installable area was 6,640 square meters. Using specifications of readily available modules (rated at 55 peak Watts [Wp] and sized 0.84 meter by 1 meter), the initial estimated photovoltaic (PV) system capacity was:

$$C_R = \left(\frac{55}{1,000}\right) \times \left(\frac{0.85 \times 6,640}{0.84 \times 1}\right) = \text{370 kilowatts-peak (kWp)}$$

According to a regional solar map,[a] the global horizontal irradiance (GHI) for Manila is at least 4.5 kilowatt-hours per square meter per day (kWh/m²/day), equal to 1,650 kWh/m². Using a typical value of 0.75 for the derate factor, ADB estimated the annual yield of the rooftop PV system to be:

$$E = 370 \times 1,650 \times 0.75 = 458,000 \text{ kWh or 458 megawatt-hours (MWh)}$$

The installed system achieved a higher yield than estimated, since the system design ended up using PV panels rated at 280 Wp, or 145 watts per square meter (W/m²). This is more than double the 65 W/m² used for the initial estimate. The higher value is also despite the roof coverage ratio being lower than initially estimated, at 60% rather than 85%.

[a] National Renewable Energy Laboratory. 2006. *Selected Asian Countries – Global Horizontal Solar Radiation*. http://en.openei.org/w/index. php?title=File:NREL-asia-glo.pdf&page=1 (accessed 20 December 2013).

Source: ADB.

1.2   Legal and Regulatory Frameworks

The lack of legal and regulatory frameworks for the promotion of solar energy, as well as the absence of strong institutions for implementing the frameworks, can prove to be major barriers in implementing the solar project. As such, this section provides a method for evaluating whether an adequate legal and regulatory framework exists.

Making that evaluation requires answering the following:

- *Are there statements of government support for renewable energy and/or solar energy?*
- *Who are the stakeholders involved in renewable energy development?*
- *How are laws and regulations enforced?*
- *What entities are legally able to develop a renewable energy project?*
- *What are the electrical and grid codes by which the project must abide?*
- *What are the building codes and local zoning laws by which the project must abide?*
- *What kind of incentives are available?*

The section below shows how to answer each of these questions. In addition, Annex 3 can be referenced for examples of actual rooftop solar policies and incentives.

Policies generally do not have the full force and effect of law. Rather, they usually serve to guide government action towards a specified goal or objective. In other words, the policy in itself is typically unenforceable against a government agency or other entities should they choose not to follow it.

Nevertheless, understanding the existing policies covering renewable and/or solar energy and taking note of how detailed and clear they are is useful. For example, policies that specifically cover solar energy development, as opposed to renewable energy development in general, may indicate how developed the solar energy market is within the country or locale. The same may be said of policies that specifically support rooftop solar PV development, as opposed to policies that only generally support solar power development or only utility-scale, ground-mounted solar PV systems. Policies are also at least useful for indicating legislature awareness and willingness to support solar development in general.

Table 2: Renewable Energy Promotion Policies in Selected Countries of Asia and the Pacific

Country	Renewable Energy Targets	Feed-in Tariff or Premium Payment	Electric Utility Quota Obligation or RPS	Capital Subsidies, Grants, and Rebates	Investment or Production Tax Credits	Reductions in Sales, Energy, CO₂, VAT, or Other Taxes	Tradable Renewable Energy Certificates	Energy Production Payments or Tax Credits	Net Metering	Public Investment, Loans, or Grants	Public Competitive Bidding or Tendering
Developed or Transition											
Australia	X	(*)		X			X	X		X	
Japan	X	X	X	X	X		X			X	
New Zealand	X								X		
Republic of Korea	X		X	X	X	X	X		X	X	
Singapore									X	X	X
Developing											
Bangladesh	X			X		X				X	
PRC	X	X	X	X		X		X		X	X
India	X	X	X	X	X	X	X	X	X	X	X
Indonesia	X	X		X	X	X		X		X	X
Malaysia	X	X	X			X				X	X
Mongolia	X	X									X
Nepal	X			X	X	X				X	X
Pakistan	X			(*)			Xᵃ			X	
Philippines	X	X	X	X	X	X		X	X	X	X
Sri Lanka	X	X	X	X		X		X	X	X	
Thailand	X	X				X				X	
Viet Nam	X			X	X		X				

CO₂ = carbon dioxide, PRC = People's Republic of China, RPS = renewable portfolio standard, VAT = value-added tax.

Notes: Entries with an asterisk (*) mean that some states or provinces within these countries have state- or province-level policies, but no national-level policy. Only enacted policies are included in the table; however, for some policies shown, implementing regulations may not yet have been developed or effective, leading to lack of implementation or impacts. Policies known to be discontinued have been omitted. Many feed-in policies are limited in scope or technology. Some policies shown may apply to other markets beside power generation, for example, solar hot water and biofuels. This table was produced based on information available in 2012 and earlier and may not capture the changes and incentives provided since then.

ᵃ This is present in policy but was not indicated in the source.

Source: Renewable Energy Policy Network for the 21st Century (REN21). 2013. *Renewables 2013 Global Status Report.* Paris: REN21 Secretariat.

1.2.1   Are there statements of government support for renewable energy and/or solar energy?

Statements of government support for renewable/solar energy may consist of policies that guide action towards renewable energy development. They are contained in a variety of places, such as national energy development plans, utility development plans, and policy statements from lawmakers or energy regulatory agencies. An example of such a policy would be a presidential decree that supports increasing renewable energy in the overall energy mix.

Like renewable/solar policies, renewable/solar energy laws guide government action towards its development. However, unlike policies, laws are issued by formal lawmaking processes and tend to be more comprehensive and contain enforceable provisions. An example would be a renewable energy act, issued by the national government's legislature, that calls for the Ministry of Finance to set up a renewable energy fund for supporting project development. Another example would be a law that creates a new government agency for the purpose of establishing a feed-in tariff and overseeing the issuance of renewable energy project permits. (See Table 2 for more examples).

That being said, renewable energy laws vary in scope and nature across jurisdictions. Some laws are very broad and serve more as policies without any enforceable provisions. However, no matter what the scope is, identifying whether there are such renewable energy laws and policies is a good first step in determining what type of government support exists for implementing a rooftop solar PV project. It will also help identify the government agencies that will be involved.

One way to start locating laws and policies is to refer to online databases kept by consulting institutions like the International Energy Agency or the International Renewable Energy Agency. They regularly track and publish information about different countries' national renewable energy laws and policies.

1.2.2   Who are the stakeholders involved in renewable energy development?

Developers should identify each government agency and institution, as well as its specific tasks stated in any national renewable energy law or their enabling regulations. Their identification can help developers to determine which agency handles what permits, as well as which entity would be most appropriate to address specific inquiries. In other words, the task here is to determine "who does what" in solar PV development.

Very often, entities will have overlapping tasks in implementing a program, overseeing budgeting, overseeing the application and licensing process, and so on. Some countries, however, have begun instituting a "one-stop shop" for obtaining all of the necessary permits and licenses for renewable energy projects. If that is not the case, common entities to seek permits and licenses from would include ministries (of energy, housing, finance, urban development, science and technology, etc.), ministerial subagencies, utilities, and electricity sector regulatory agencies.

1.2.3   How are laws and regulations enforced?

Based on a renewable energy law, or in lieu of such a law, a government entity may be specifically tasked with implementing a country's renewable energy (and solar) policy and issuing regulations. Depending on the jurisdiction, laws and regulations may have enforcement mechanisms. They often take the form of fines or other penalties for noncompliance.

Not only are enforcement provisions important for the developers, they indicate how seriously an entity takes the development process—if such penalties have been imposed for default. For instance, many jurisdictions have anticorruption laws with agencies dedicated to receive complaints and handle them accordingly.

1.2.4 What entities are legally able to develop a renewable energy project?

In general, the project developer and contractors must be individuals or legal entities that are registered and authorized to do business in the country in accordance with corporate laws. Sometimes, foreign ownership of a corporation is an issue.

1.2.5 What are the electrical and grid codes by which the project must abide?

After identifying the agencies that issue and enforce relevant electrical and grid codes, developers need to learn about the codes that affect solar PV installations. There are likely to be specific codes for solar PV and rooftop solar PV systems, such as safety, installation, and maximum voltage.

1.2.6 What are the building codes and local zoning laws by which the project must abide?

Further, developers need to identify national and local building codes. Local districts may impose zoning restrictions that would impact the design of a rooftop solar installation. For example, there may be a local zoning law that requires solar panels not to be visible from the vantage point of a person standing at ground level on any public street.

1.2.7 What kind of incentives are available?

This section describes common fiscal policy incentives for renewable energy development. They include renewable portfolio standards, renewable energy certificates, carbon credits, grants and rebates, low-interest loan programs, tax incentives, feed-in tariffs, and net metering.

How the government implements these incentives and makes them available to developers depends on the jurisdiction and type of incentive. Thus, after identifying whether one of the following incentives exist, developers should make further inquiries to the relevant implementing agencies for determining the process for obtaining the incentive.

Renewable Portfolio Standard
The renewable portfolio standard (RPS), also known as a renewable obligation or quota system, requires certain obligated entities to source a specified amount or percentage of their electricity from designated renewable resources. They are commonly applied to electricity utilities (e.g., distribution companies), but could apply to certain customer categories, like large commercial or industrial customers. Obligated entities that do not meet their specified renewable energy target face penalties.

In some cases, the RPS may specifically require that solar resources meet a minimum share of a larger renewable energy target (referred to as solar "carve-outs").

Renewable Energy Certificates
Renewable energy certificates (RECs) are a market mechanism used in combination with the RPS. They represent property rights to electricity generated by renewable energy. For instance, one REC may be issued for each megawatt-hour of electricity produced by renewable energy. RECs then become a tradable instrument—a qualified renewable energy generator may sell RECs to entities that have RPS targets.

Clean Development Mechanism
The Clean Development Mechanism (CDM) is a type of carbon credit developed under the Kyoto Protocol, issued under the United Nations Framework Convention on Climate Change (UNFCCC). Once created, a carbon emission reduction credit (CER) operates as a tradable instrument, with each CER representing a ton of carbon dioxide

abated annually. Upon earning credits, project developers may sell them to industrialized countries for meeting their emission reduction targets under the Kyoto Protocol.

Industrialized countries, as defined under the Kyoto Protocol, agree to certain emission reduction targets. To comply with those reductions, industrialized countries may purchase CERs from qualifying projects in developing countries.[9]

The capacity of a rooftop solar PV project is generally small and a simplified methodology is in place for registering the CERs. However, rooftop solar PV projects may qualify for receiving carbon credits under UNFCCC standards. It generally requires meeting two criteria: (i) projects must result in reducing measurable greenhouse gas emissions, which would not have occurred without this project; and (ii) projects must demonstrate their contribution to their country's environmental and sustainable development goals.

The process of obtaining carbon credits first involves public registration and approval with the "designated national authority." The UNFCCC maintains a searchable database of these authorities on its website.[10] After successful registration, an independent body verifies the carbon offsets and tracks the sale and trade of carbon credits.

The UNFCCC's website provides all the information on how to qualify for carbon credits and participate in their trade. The designated national authority should also be able to provide information and assistance.

Grants and Rebates
A national or local government may offer grants and rebates for the development of rooftop PV. These are cash incentives intended to help finance part of the capital needed for installing a PV system by reducing overall project costs. Capital grants provide a sum of money upfront, while rebates provide it after successful installation, or after the system is successfully connected to the grid. Who and how to receive payment will depend on local regulations. They may require a solar contractor/installer to apply for grants or rebates, and then apply them in full to the cost of the system to reduce the customers' out-of-pocket expenses. Alternatively, the customer may apply for grants and rebates directly.

Low-Interest Loan Programs
Governments or electric utilities can offer low-interest loan programs or revolving loan funds to support solar installations and may leverage funds through cooperative arrangements with banks and other private sources of finance. Alternatively, public authorities may provide information or technical assistance to homeowners or businesses that seek to form a cooperative, with the intent of negotiating a competitive bulk purchase of solar power from available providers.

Tax Incentives
Tax incentives would reduce the overall taxes paid in relation to solar PV development. Typical tax incentives for solar PV development include accelerated depreciation on capital and/or equipment, property tax abatement for green buildings, exemptions, holidays, and credits.

They also include tax exemptions, which may be available for the sale or importation of solar PV-related equipment, or in relation to operation costs.

---

[9] See Kyoto Protocol, Article 12. Available at the United Nations Framework Convention on Climate Change website at http://cdm.unfccc.int/about/index.html
[10] See United Nations Framework Convention on Climate Change, "What is a designated national authority?" Available at http://cdm.unfccc.int/DNA/index.html

Annex 3 can be referenced for examples of tax incentives and exemptions.

Feed-In Tariffs
Feed-in tariffs (FITs) are long-term purchase agreements for renewable energy electricity. Payment is typically based on the amount of electricity delivered to the grid, per kilowatt-hour.

FITs vary widely across jurisdictions. Common variables for whether a project qualifies include technology type, project capacity, resource quality, specific location, or the type of fuel it displaces. Another variable is the length of time a project is eligible for receiving a FIT, which ranges anywhere from 5 to 25 years (Couture et al. 2010).

Payout schemes also vary. The two most common types are the fixed-price FIT and the premium FIT. The *fixed-price FIT* guarantees a fixed rate for every kilowatt-hour delivered to the grid. The rate is independent of the market price of electricity. The *premium FIT* operates as a premium price paid out in addition to the spot market electricity price. The rate of the premium may be fixed, or may be designed to vary according to the spot market electricity price. The premium FIT is less commonly deployed, compared to the fixed-price FIT.

To be eligible for a FIT, registration with the implementing agency may be necessary. FIT rates can often be found on ministerial or other government agency websites.

Net Metering
Net metering allows for generators of solar PV electricity to be compensated for the electricity they do not use and feed it into the grid.

Compensation is usually in form of a credit for the excess electricity (measured in kilowatt-hour) against the generator's electricity consumption from the utility. By implication, the price of electricity generated on the roof becomes the same as the retail electricity tariff, which is usually much higher than the wholesale electricity purchased by the distribution utility. However, this is often not the case—in many instances, PV owners sell at wholesale.

Unused kilowatt-hour credits usually expire after a defined time period (e.g., 1 year). However, in some cases, the utility will pay a fixed price to the customer for a limited number of unused credits.

If employed successfully, net metering is a useful alternative to having batteries store electricity for use when the solar energy system is not generating sufficient electricity.

1.3   Permits and Licensing

Acquiring all the necessary permits and licenses can prove to be a daunting task, especially in jurisdictions that have not yet implemented "one-stop shops" for obtaining them. In such a case, a developer may need to obtain them from multiple agencies, with each having their own application rules, application fees, and processing times.

This section provides a list of typical licenses and permits associated with rooftop solar PV projects. Annex 2 provides an example of the permits and licenses ADB obtained for its ADB Headquarters Rooftop Solar Project.

Typical licenses and permits include the following:

(i)   renewable energy developer registration;
(ii)   zoning clearance;
(iii)   building permit or building renovation permit;

(iv) electrical permit;
(v) mechanical permit;
(vi) structural permit;
(vii) fire permit;
(viii) environmental impact assessment, or certificate of exclusion thereof;
(ix) renewable energy generation and/or sale license;
(x) clearance or permit for grid interconnection;
(xi) helipad height clearance permit; and
(xii) activity clearance for transport of materials.

Government websites are useful in ascertaining which agency is in charge of what permit, as well as its requirements. However, directly contacting relevant agencies to obtain up-to-date information is strongly encouraged. This step recognizes that many entities' websites are not up to date, and changes (or anticipated changes) to permit and licensing regulations may not always be published.

1.4 Financing Options

A system can either be purchased and financed directly or through a third party (via solar leasing or power purchase agreement).

Both options require estimating the net investment cost as a first step. This can be achieved using readily available information, such as indicative quotes or values for similar projects. If a market for PV systems has not been established in the country, it may be necessary to obtain proxy costs from markets in other countries.

The net investment cost is the cost of the system minus any cost offsets. The cost of the system includes the components and labor for design and installation, operation and maintenance, interest on loans, plus the cost of any required permits. Cost offsets would include tax credits, incentives or rebates, annual electricity savings, and any revenue from leasing the roof.

Box 5 provides an example of how to calculate the net investment cost, using ADB's experience in estimating the cost for its ADB Headquarters Rooftop Solar Power Project.

Box 5: ADB Solar Power Project Cost and Price Estimate

Using price quotes from suppliers, the Asian Development Bank (ADB) estimated the project would cost $1,300,000 for a solar photovoltaic (PV) installation with 350–400 kilowatts-peak capacity and an annual output of 458 megawatt-hours. The preliminary financial model assumed that the project cost would be covered in three parts: (i) a $700,000 ADB private sector loan with a 15-year term and 4% annual interest, (ii) a $210,000 commercial loan with a 7-year term and 6% interest rate, and (iii) a $390,000 equity of the developer with a return of 15%.

The operation and maintenance expenditure and insurance was assumed to be 1.0% of the project cost, and it was assumed there would be no tax liability (on capital expenditure or profit).

In determining price, ADB sought a positive net present value over the 15-year term of the power purchase agreement with a discount rate of 10%. ADB thus determined that the levelized electricity price would need to be $0.351 per kilowatt-hour. This price level was predicted to be sustainable—the developer would get a return of 7.8% in the first year, which would increase if the developer repaid the commercial debt in the first 7 years. Thereafter, the return on equity would be well over 20%.

Source: ADB.

1.4.1 Direct Purchase

With a direct purchase, the project developer and/or building owner uses their own funds or obtains debt financing for the rooftop solar PV project.

One advantage of this financing scheme is that the developer reaps all financial benefits stemming from the project. However, the developer is likewise responsible for all costs and risks, which can be significant when the developer lacks expertise of managing development of solar PV systems. Further, capital investments would require a more rigorous internal review and approval process in any company because of the impact on the balance sheet.

1.4.2 Third-Party Financing: Solar Leasing Arrangements and Power Purchase Agreements

Under third-party financing mechanisms, the owner does not purchase the PV system but instead enters into an arrangement with a company to make periodic lease payments or pay the third party for the electricity produced by the system. Third-party financing therefore facilitates a building owner's access to competitively-priced solar electricity, without the burden of significant up-front costs of PV systems.

One disadvantage of direct versus third-party financing is the additional overhead cost involved. It at least requires drafting a suitable third-party agreement, and possibly working with public authorities or regulatory bodies to ensure that an adequate legal framework exists to support their use (US DOE 2011).

The advantage of third-party financing is that the building owner could reduce the risk and complexity involved in owning and operating a PV system, provided that the third party has expertise and is specialized in the business. Third parties also may have methods to spread the risks and costs over a larger number of customer-based installations.

There are two main types of third-party financing mechanisms. These are solar leasing arrangements (SLAs) and power purchase agreements (PPAs). They differ by the type of payment the third party receives for building and maintaining the rooftop solar PV system. Under SLAs, the building owner pays the third party back in monthly installments, which are *fixed and independent of the electricity actually produced.* Under PPAs, the building owner *pays per unit of electricity generated.* The electricity rate may be fixed or increase or decrease at a specified rate over time (Kollins, Speer, and Cory 2010).

For building owners, where both SLA and PPA financing models are available, there are a number of factors to consider when comparing the alternative approaches and their merits. These include the up-front cost of each option, the rate at which either the lease payment or the PPA price increases each year, whether maintenance is included in the lease, whether online system monitoring is included, and the costs associated with terminating the lease or the PPA during the term of the agreement.

Another factor to consider is the fixed payment scheme in SLA. For an SLA to be economical, the monthly leasing payments should be no more than the monthly electric bill savings (Kollins, Speer, and Cory 2010). The customer therefore takes more risk with a solar lease than for a PPA since the lease payment is fixed and independent of electricity produced, and the building owner may also be responsible for property insurance (Kollins, Speer, and Cory 2010).

While PPAs seem more flexible, not all markets allow for PPAs. They usually either allow for an SLA or PPA, and therefore no choice as to either one. This is due to a number of factors, including whether or not the PPA is permissible by law and how electricity sales are taxed vis-à-vis leases.

ADB's rooftop project adopts the PPA approach. An explanation of its PPA modeling scheme appears in Box 6.

Box 6: ADB Rooftop Solar Project Business Model

The Asian Development Bank (ADB) rooftop solar system uses the power purchase agreement (PPA) business model, which minimizes technology risk to ADB. The vendor designs the system, leases the roof space from ADB, purchases and installs the equipment, and then owns and operates the facility for 15 years. All the electricity generated by the rooftop photovoltaic (PV) system is sold to ADB at a fixed price. The vendor guarantees the minimum amount of electricity to be generated and sold during the term of the PPA. The complete PV system will be transferred to ADB at the end of the 15-year agreement at no cost.

A PPA was prepared (see sample in Annex 9) with provisions to sell all the generated power to ADB at a final bid price (in dollar per kilowatt-hour) determined through a competitive bidding process. Because payment is tied to electricity delivered to ADB, payments are lower when less electricity is delivered. If the solar radiation is less than anticipated for any reason, or the efficiency of the PV system is lower than the design value, the power provider bears the risk.

In addition to purchase price, the PPA stipulates a minimum amount of electricity output (in kilowatt-hour) that will be produced each year. The contractor guarantees that he will provide at least 90% of that expected output. If the electricity output is lower than 90% of expected output, the contractor pays to ADB the difference between the contracted purchase price and the price ADB would pay its other electricity provider(s) to make up for that difference. In the event that the guaranteed energy output is not achieved, ADB would also have the option to purchase the facility at half of the value of the system. This cumulative output guarantee provides an additional incentive to ensure that the plant operates close to design capacity.

To protect the contractor, the expected output is calculated cumulatively after every 3-year period of operation to allow for inter-annual variations in weather. It can also be adjusted for operating efficiencies being lower than expected, such as where the design capacity is overstated. Also, in the event that the electricity output is higher than expected, ADB purchases the excess electricity at the same price of $0.379 per kilowatt-hour.

Source: ADB.

Chapter 2: System Design

This section describes the components of a rooftop solar PV system, and then addresses the following important considerations for its development:

(i)   on- or off-grid option,
(ii)   site characterization and assessment,
(iii)   solar resource assessment,
(iv)   shading analysis,
(v)   array configuration,
(vi)   solar PV module selection,
(vii)   mounting system design,
(viii) inverter selection,
(ix)   wiring design,
(x)   system performance assessment, and
(xi)   due diligence.

These considerations are important because they have a direct impact on a system's rated capacity.

System design and procurement can be considered concurrently, as the project developer's experience in designing the system would be invaluable toward implementing the project successfully. ADB included its system design requirements in its bidding documents, which required bidders to design their proposed solar systems based on those requirements.

Those bidding documents are available upon request from ADB.

2.1   The Components of a Rooftop Solar Photovoltaic System

A "rooftop solar PV system" is the name for the whole set of components in a solar PV power plant that collectively supply usable electric power. This is composed of the PV panels and the balance of system (Figure 3), which includes all other components of the system.

The balance of system has essential components that must be included in all systems. These components are mounting frames to secure the modules in place on the roof; inverters to convert the direct current (DC) output of the modules to alternating current (AC); and electrical wiring, switches, and interconnections.

Optional components include metering equipment to indicate energy usage and system performance, batteries to provide energy storage or backup power in case of a power interruption or outage on the grid, solar trackers, weather data collecting instruments, and performance monitoring software (US DOE 2006).

Figure 2 illustrates how these components fit together. The figure shows both essential and optional components.

Figure 2:   Diagram of a Solar Photovoltaic System

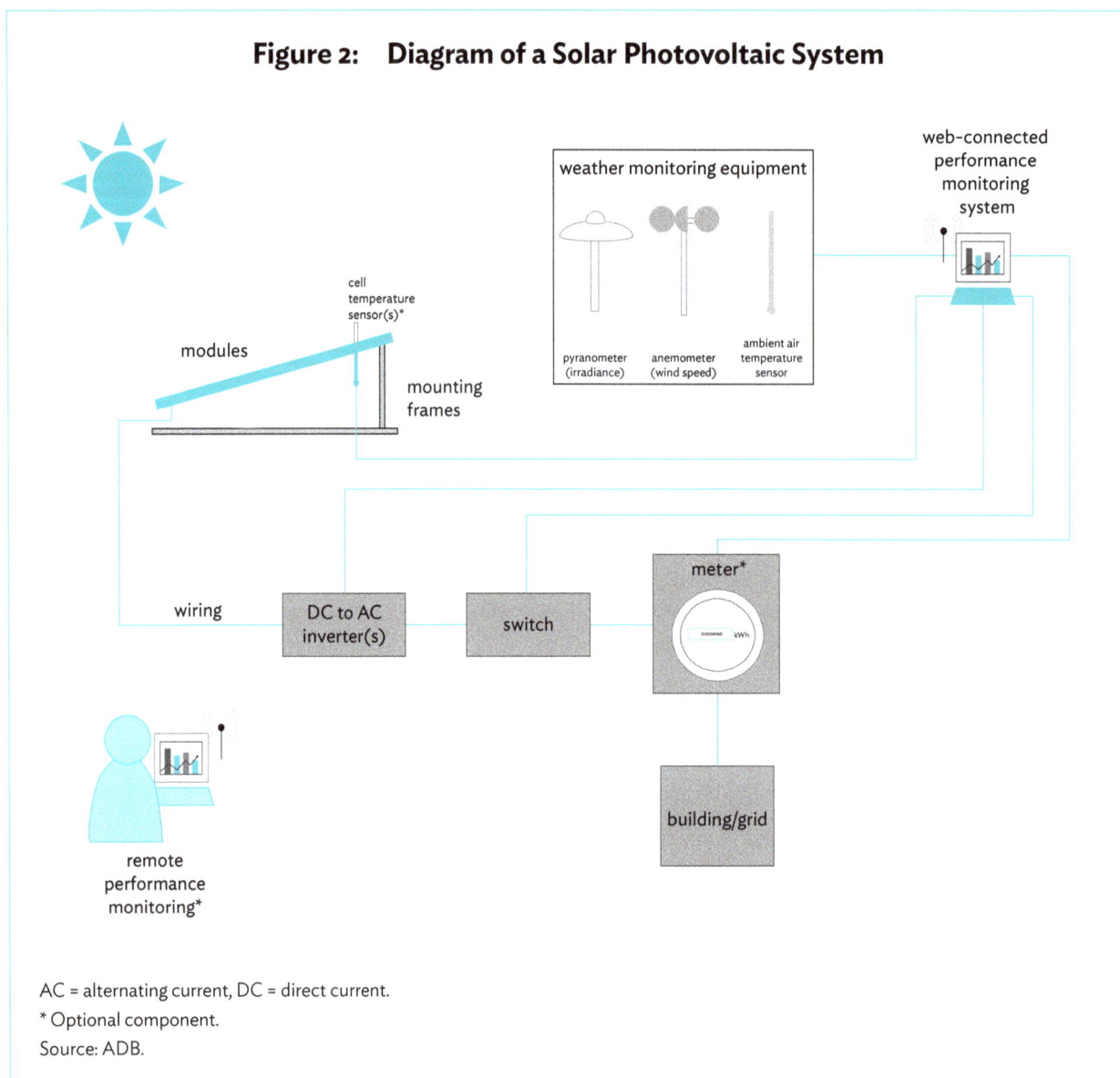

AC = alternating current, DC = direct current.
* Optional component.
Source: ADB.

2.2   On- or Off-Grid Option

PV systems are either on-grid or off-grid. On-grid means connected to the electricity grid, and off-grid means independent from the grid (i.e., stand alone).

In *off-grid systems,* the solar panels feed DC electricity to batteries, and an inverter converts stored battery energy into AC electricity on demand. The batteries allow the system to provide power when the sun is not shining. Such an arrangement is suitable where a grid connection is not possible and when the solar system is able to generate more electricity than the building's demand during part of the day.

In *on-grid systems,* an inverter converts the DC output of the panels to AC electricity that synchronizes with the grid electricity. Batteries are not needed to store the electricity because the excess, if any, is supplied to the grid.

On-grid systems are suitable in areas with an existing grid connection and are essential when the solar system will only produce a portion of the demand. This was the case for the ADB rooftop solar system (Box 7).

One advantage with on-grid systems is being able to avail of net-metering benefits, if any. As explained previously, when a PV system is able to meet the total electricity demand of the building, excess power from the grid-tied system can be sent directly back into the grid. Thus, where net metering laws are in effect, a special utility meter will measure the power coming in from the utility and the surplus power flowing out from the solar PV system, enabling site owners to act as micro-generators and be financially rewarded for any net electricity generation they provide.

On-grid systems may also include battery backup or uninterruptible power supply systems that can operate selected circuits in a building for hours or days during a utility outage. In these hybrid systems, the solar panels feed electricity through a battery to the inverter and supply AC electricity when needed. When the batteries are full, the electricity produced by the solar plant will feed to the grid.

Box 7: ADB Rooftop Solar Power Generation System

The Asian Development Bank (ADB) connects its rooftop photovoltaic-generated electricity to ADB substations, where the electricity combines with electricity drawn from the grid before it is supplied to the headquarters.

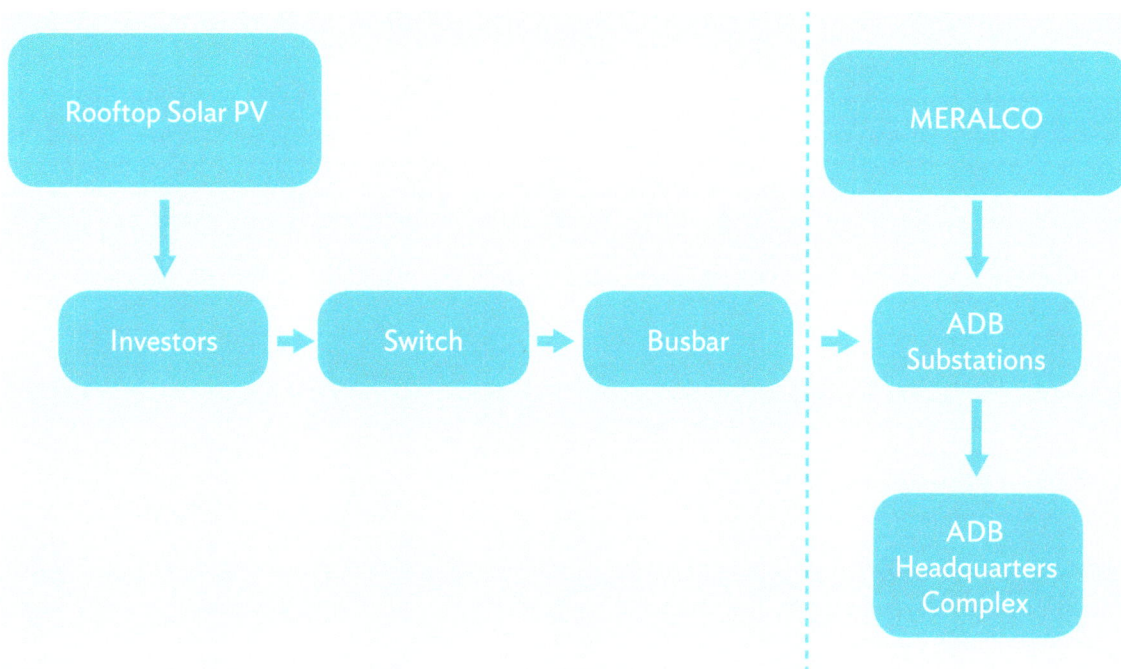

ADB = Asian Development Bank, Meralco = Manila Electric Company, PV = photovoltaic.
Source: ADB.

2.3 Site Characterization and Assessment

A detailed site characterization provides the necessary information needed for the design of the PV system.

The first part of the site assessment is to characterize the following physical conditions of the project site:

(i)   roof latitude, longitude, elevation, and orientation;
(ii)   floor plan of roof deck, with dimensions and total area;
(iii)   access to roof for installation and maintenance;
(iv)   competing uses of the roof, and proximity restrictions of the solar array from these uses;
(v)   roof plan showing area usable for the solar installation, total calculated usable area, and slope;
(vi)   roofing materials, possible mounting connection points, and restrictions to mounting, to aid with selection of mounting method (mounting systems are further discussed later);
(vii)   shading analysis, by structures on the roof and nearby structures (detailed later);
(viii)   any restrictions for aesthetic purposes;
(ix)   specifications of the strength of the mounting system required to:
    (a)   comply with building or electrical codes for solar panels, and
    (b)   meet wind speed (jet blast) rating requirements for helipads, if present; and
(x)   structural plans for the building and load capacity of the roof.

The last point is the most crucial. The roof must be capable of supporting the additional weight of the solar PV system, plus withstand maximum wind loads. It is especially important in locations such as Manila, where the occurrence of tropical typhoons is relatively common. For example, a strong wind could create an uplift force of over 244 kilograms per square meter (50 pounds per square foot) (NABCEP 2005).

Structural or building engineers will need to ascertain if the roof can support the proposed solar system, and experts may conduct a structural and seismic analysis of the final design. Some mounting systems, such as the one ADB used, come with wind-deflecting components that reduce this force. Other upgrades may be necessary to ensure structural support is sufficient.

It is also necessary to clearly define the load characterization, and wiring and interconnection with the grid, as follows:

(i)   Load characterization
    (a)   average annual electricity consumption,
    (b)   typical load profile during working days and weekends, and
    (c)   minimum demand.
(ii)   Wiring and interconnection with the grid
    (a)   interconnection location, current-carrying capacity of contacts (in switches), and electrical specifications;
    (b)   technical specifications of substation and switch gear; and
    (c)   site plan showing locations of substations, conduits, and interconnections, including space available for installing transformers.

These conditions will help with determining whether the existing electric service can accommodate the PV system. For instance, if the PV system delivers electricity to a building via a breaker in the existing service panel, the breaker size (as determined by the maximum current output from the PV inverter) would be limited to the rating of the service panel. Older buildings may require upgrades to their existing electric service in order to accommodate a new PV system.

2.4 Solar Resource Assessment

A solar resource assessment evaluates the output of a PV system, which is nearly directly proportional to the total solar irradiance incident on the system. It includes both radiation directly from the sun and downward radiation from the sky and clouds on to the system. The latter allows for systems to potentially generate electricity even on cloudy days.

At the top of the atmosphere, the total solar irradiance averages 1,367 watts per square meter (W/m²) (Stoffel et al. 2010).[11] As solar radiation passes through the atmosphere, much of this energy is absorbed or reflected back to outer space by atmospheric haze, clouds, and water vapor, leaving a global annual average of just 170 W/m² reaching the surface (World Energy Council 2007). However, many regions receive much more: most of Southeast Asia receives an annual average of 180–230 W/m² (Figure 2).

Solar radiation is measured at ground level by an instrument known as a pyranometer. When mounted horizontally, the instrument measures global horizontal irradiance (GHI). This irradiance is the geometric sum of two components:

- direct normal irradiance (DNI), which is the radiation directly from the sun; and
- diffuse horizontal irradiance (DHI), which is the total radiation reflected off of the clouds and other molecules and particles in the atmosphere:

$$GHI = DHI + [DNI \times cos\ (z)]$$

The solar zenith angle, z, is the angle between the direction of the sun and the zenith (Figure 3).

Although light direction is irrelevant to a photovoltaic cell, diffuse light is less affected by shading than direct light, so it is important to measure data on both global horizontal irradiance and the fraction of light that is diffuse.

Correct use of appropriate irradiation data is paramount for accurately estimating how much electricity a solar PV system will generate. Poor quality solar irradiation data for a site location, even when used with excellent simulation programs, will yield inaccurate results when compared with actual system performance.

Ideally, GHI data should be accurate to within –1% and +3% of the World Radiometric Reference, which is the internationally accepted reference for ground-based measurement of solar radiation (Stoffel et al. 2010). Data should also ideally be of long enough duration to capture the majority of interannual variability. This means a data record of at least 10 years' duration.[12]

Long-term data of at least 10 years' duration is difficult to come by, since the specified accuracies must be obtained by high-quality surface measurements that are rarely available for such a long time period. However, an alternative exists and it should be sufficient enough: long-term solar data derived from weather satellite imagery can be merged with any short-term, high-quality measured data available from the site. However, the type of short-term data used will depend on the size of the project.

---

[11] Total solar irradiance at the top of the atmosphere varies from 1,321 W/m² to 1,415 W/m² throughout the year due to Earth's elliptical orbit around the sun.

[12] Solar radiation varies from year to year, with the amount of variation dependent on general climatic conditions. For example, interannual variation of GHI in Australia was found to be ±8% from the long-term mean, with tropical coastal locations having significantly higher variation than semi-arid inland locations (Blacker and Hidalgo Lopez 2011). In the United States, GHI interannual variability is typically 8%–10%, while interannual variation of DNI is roughly double that of GHI (Stoffel et al. 2010).

Figure 3: Global Horizontal Irradiance as a Combination of Direct Normal Irradiance and Diffuse Horizontal Irradiance

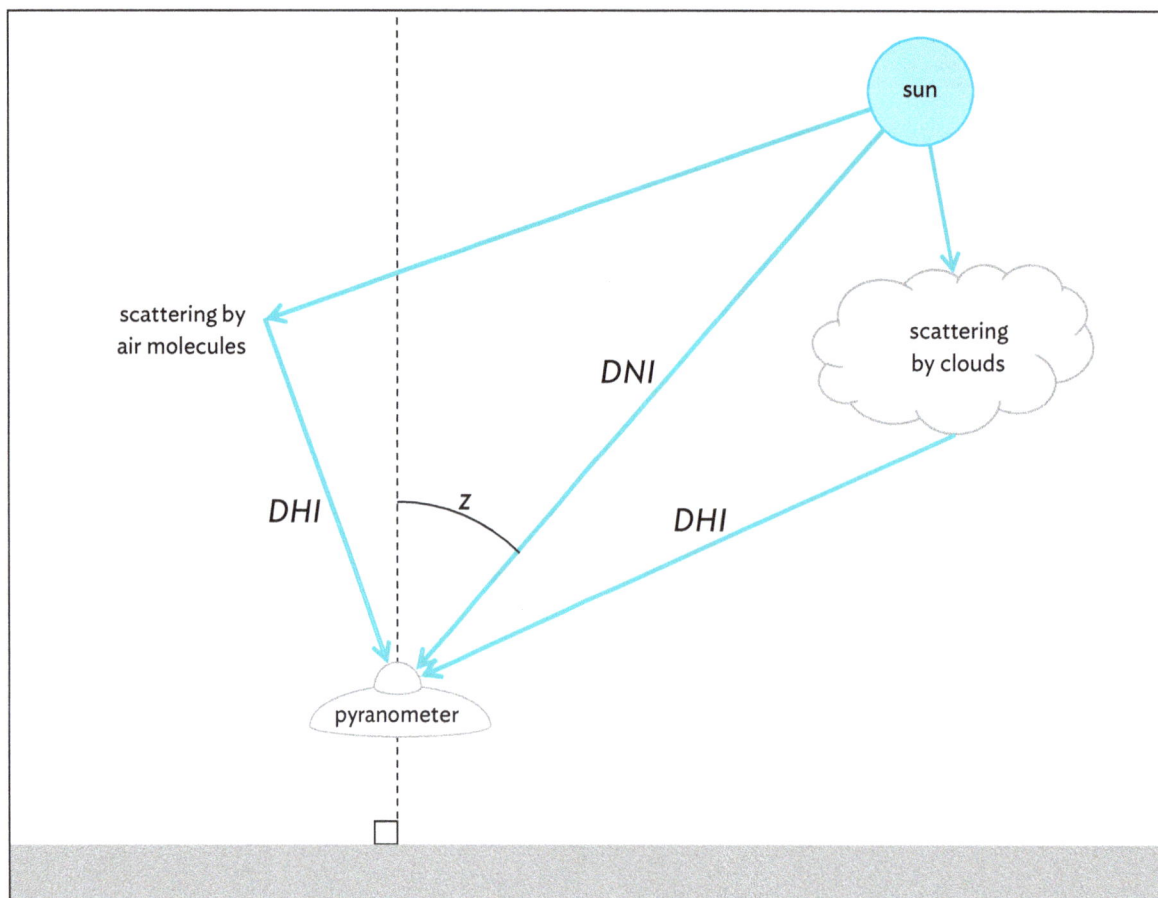

DHI = diffuse horizontal irradiance, DNI = direct normal irradiance, z = solar zenith angle.
Source: ADB (adapted from Stoffel et al. 2010).

For *small projects* (<100 kilowatts), satellite-derived data used with simulation programs may be sufficient to estimate system output.

For *medium-sized projects,* such as the ADB Rooftop Solar Power Project, it is advisable to obtain at least a short period of high-quality ground measurements and estimate the interannual variability from the long-term satellite data. The ground measurements should be used to remove any bias from the satellite data. Box 8 describes the initial resource assessment undertaken for the ADB Rooftop Solar Power Project.

An alternative to obtaining ground measurements is to obtain data from the local meteorological agency if a station is located close to the site. These data are often long-term, high-quality, and publicly available. However, it should still serve only as an alternative because there may still be uncertainty in the estimation. The uncertainty would be due to there being insufficient data against which satellite data can be properly correlated.

For *large projects,* collecting high-quality, on-site data for at least 1 year may be necessary. The data would be collected using a well-calibrated and maintained pyranometer in order to be of sufficient accuracy. Should the PV panels be tilted at roughly the same angle as the latitude of the site so that they point more normally toward the sun, a pyranometer tilted at the same angle might also be deployed.

A manual developed by the United States National Renewable Energy Laboratory provides detailed information on data collection procedures (Stoffel et al. 2010). While this method will provide a more accurate data set, it is more costly and time-consuming. Also, even if data are collected at the site for a full year, they should still be correlated with a longer-term data set to account for interannual variability.

Bidders were advised to conduct their own, more detailed resource assessments based on these data and any additional data. An independent evaluation supplemented and confirmed the solar resource using a satellite-based surface solar energy dataset, a map generated using satellite data and solar energy modeling, and a software tool that estimates solar resource for a given location.[13] These datasets showed that the Philippines has an annual potential global horizontal irradiance averaging 5.1 kilowatt-hours per square meter a day, although considerable variability can occur across the region due to the effects of localized terrain influences on cloud formation.

Box 8: Resource Assessment for the ADB Rooftop Solar Power Project

The Asian Development Bank (ADB) acquired 2 years of hourly average solar resource data from the Philippine Atmospheric, Geophysical and Astronomical Services Administration (PAGASA), as collected in Quezon City (Science Garden). Its location is 6.5 kilometers from ADB headquarters.

The following table is a summary of the data that ADB provided to bidders in the project technical outline:

Solar Radiation at PAGASA Science Garden in 2007

Month	GHI (kWh/m²/month)	DHI (kWh/m²/month)
January	116.2	69.33
February	125.8	66.18
March	170.0	76.14
April	176.2	75.60
May	171.4	74.18
June	136.0	77.85
July	144.7	83.19
August	111.8	74.06
September	127.5	70.92
October	114.1	69.23
November	100.4	65.02
December	106.8	61.39
TOTAL (Annual)	**1,600.9**	**863.10**

DHI = diffuse horizontal irradiance, GHI = global horizontal irradiance, kWh = kilowatt-hour, m² = square meter.
Source: ADB.

[13] Surface Solar Energy Data Set (National Aeronautics and Space Administration, Meteonorm, and the Climatological Solar Radiation [CSR] Model [NREL]).

2.5   Shading Analysis

It is vital to understand the shading on the rooftop, particularly in urban areas where surrounding buildings and structures can cast shadows on the roof. Even a small portion of shading on a PV array can significantly reduce output because of how the panels and array are electrically configured.

Ideally, all portions of the roof should be unshaded for at least 6 hours a day, preferably between 9 a.m. and 3 p.m. The shading analysis has to be carried out for all sunshine hours throughout the year to account for the seasonal variation of the sun path. This helps in the selection of the best location to mount the solar modules and gives a more accurate estimation of the annual output of the PV system.

Shading analysis methods range from simple manual methods to more complex three-dimensional (3D) rendering. This handbook explains both methods.

The *manual method* will give a quick impression of the solar window (aperture) of the sky, immediately showing the structures that block the site from collecting full solar radiation. To apply this method, capture an image of the 360-degree view of the sky from the roof using an instrument such as the Solar Pathfinder (a reflective dome that also includes a chart of the solar path) or using a camera with a fish-eye lens pointed vertically. Capture the sky view from various locations on the roof, measuring the location and direction at each position with a compass and GPS

Figure 4:   A Spherical Image from the ADB Rooftop Overlaid on a Solar Chart

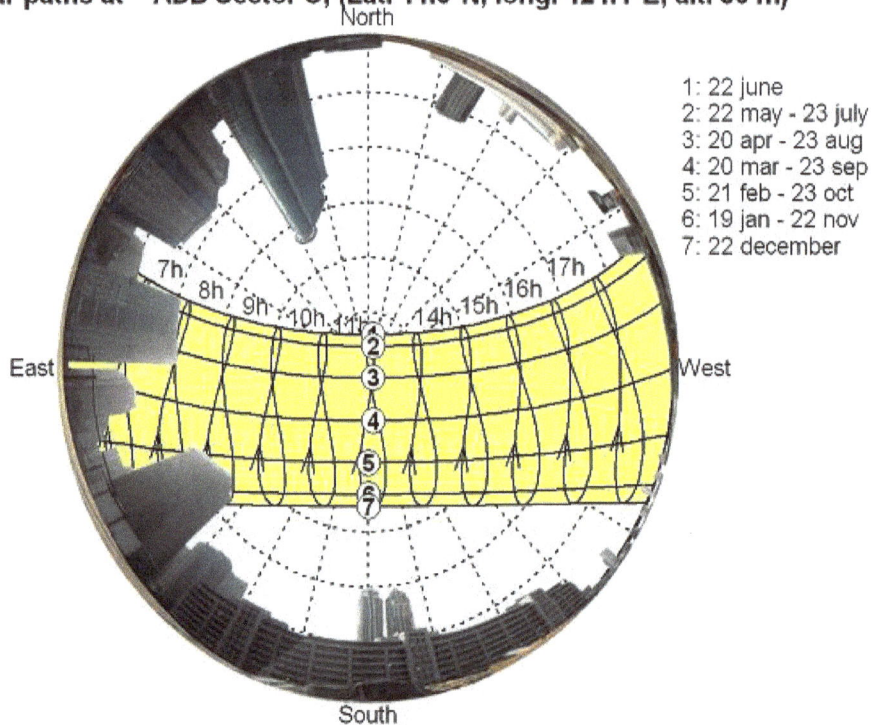

Note: ADB's Headquarters is located on the northern hemisphere. In a solar path chart of this type, the sun's position is taken from the perspective of an observer facing the equator. Thus, east appears on the left and west appears on the right.

Source: ADB.

device. Overlay the photograph on a sun path diagram for latitude and longitude to see the times of day at various times of the year at which that particular roof location will be shaded, as in Figure 4. Instruments will typically already include this overlay and often also include software to analyze the solar window. This method will only capture shading from existing structures.

Second, the *3D rendering method* requires computer software to conduct the shading analysis. To apply this method, gather location, orientation, and dimensions (length, width, and height) of surrounding structures using satellite images and/or a ground survey. Estimate heights of buildings where unknown either by multiplying the number of floors by 3 meters per floor, or by comparing the length of the building shadow in a satellite image with a building with a known height. The software may also allow addition of existing roof structures such as chimneys; heating, ventilation, and air conditioning (HVAC) units; chillers; vents; ducts; and parapets. The computer simulation should output the loss due to shading on predefined sections of the roof.

Getting information about possible high-rise buildings that may be built in the future is also important. High-rise buildings may cast shadows on the PV array, both within the area owned by the building owner and outside, especially in areas where future development is likely. Information on future construction can be discerned from looking at zoning laws (especially height restrictions), planning permissions or building permit approvals for the surrounding areas, and vacant lots. Future nearby structures can easily be added into the shading model, and may also be added to the manual method.

In addition to shading from surrounding structures, shading from the array itself also needs to be considered. This is discussed in the next section.

Box 9: Shading Analysis for the ADB Rooftop System

The Asian Development Bank (ADB) installed its rooftop solar system on top of a low-rise block with a number of taller buildings adjacent to it. Therefore, ADB found it necessary to conduct a comprehensive shading analysis, which consisted of both the manual method and the three-dimensional rendering method (**Annex 4**). The resulting shading analysis became a very critical part of the assessment of potential generation and essential input to the system design. ADB provided bidders with spherical photographs taken from the roof, and required bidders to verify ADB's shading analysis by carrying out their own.

Source: ADB.

2.6 Array Configuration

The array configuration analyzes the PV module tilt angle, orientation, and spacing. The purpose of the analysis is to estimate their impact on system performance and visual appearance of the system.

Tilt angle. Fixed panels will generally collect the maximum solar irradiation per area if tilted in the direction of the equator at an angle equal to the latitude. Arrays that deviate from this ideal tilt and orientation will generally collect a lower amount of irradiance, but may still be feasible. For instance, an array installed at a latitude of 40 degrees North and with a tilt angle of 30 degrees still receives 94% of the annual solar radiation if facing southwest instead of due south, and receives 81% of the annual solar radiation if facing due east.

At lower latitudes, the tilting and orientation of the array will have less of an impact than on systems at higher latitudes (NABCEP 2005). Nevertheless, an optimal system configuration should be considered for large systems even in these lower latitudes. Installations at low latitudes should still have a tilt angle of at least 10 degrees to allow for self-cleaning of the modules through rainfall. Higher tilt angles do, however, result in higher wind load.

Orientation. The orientation can be optimized by understanding the typical diurnal cloud conditions and the time of maximum load that the array is serving. For example, in sites such as Manila, clouds tend to be greater in the afternoon than in the morning, so that tilting the array somewhat east of south will likely receive a higher resource during the morning hours. However, if the maximum load is in the afternoon, it may be better to tilt the array somewhat to the west of south to maximize solar input during the afternoon. Even though there may be more clouds, the resource received by the array can be maximized at a time that coincides with the maximum load.

Spacing. Spacing refers to the space between panels and potential inter-row shading. Inter-row shading can seriously impact system performance, as just a 6-inch (15 centimeter) shadow could disrupt energy production from an entire string of modules.

As a general rule, the space between modules should be at least three times the height of the tilted modules at higher latitudes, and at least two times the height at lower latitudes (NABCEP 2005). In adapting this rule to a rooftop system, there is limited space so it is necessary to balance the number of panels that can fit on the roof with the losses from shading. One way to balance these two factors is by installing the PV modules at lower tilt angles. Doing so reduces the spacing necessary to avoid inter-row shading, therefore allowing more modules to be installed in the available roof space.

Aesthetics. The final consideration is how the array configuration will affect the system's aesthetics. One example is that building owners with sloped roofs tend to prefer the aesthetics of solar panels installed parallel to the roof.

Box 10 describes the array configuration design for ADB's rooftop solar system.

Box 10:   ADB Rooftop Array Configuration

The independent body that conducted the initial due diligence of the system design (**Section 2.11**) recommended that the photovoltaic (PV) modules for the Asian Development Bank rooftop solar system be tilted toward the south at angles close to the latitude of the location (14.5 degrees) to optimize the annual solar energy captured by the PV system. They also recommended to space module rows by at least 0.7 meters to minimize internal shading losses.

Finally, the system was installed facing due south with a tilt angle of 12 degrees and an inter-row spacing of 0.28 meters. This configuration allows for the array to collect the maximum amount of sunlight given the local shading and climatic conditions. Having a slightly smaller tilt angle than the latitude also allows the rows to be closer together, meaning that more modules fit on the roof, resulting in a higher array capacity.

Source: ADB.

2.7   Solar Photovoltaic Module Selection

This section seeks to illuminate how solar PV module technology works, to select the best technology for the project's purpose and/or building owner's needs. The questions that this section will answer below are:

- *How do photovoltaic cells work?*
- *What is the difference between a solar or photovoltaic cell, module, and panel?*
- *What are the types of solar modules, and which one is best?*

2.7.1　How do photovoltaic cells work?

A photovoltaic cell contains two or more semiconducting materials in close contact with each other. These semiconductors convert light directly into electricity using the photovoltaic effect.

The photovoltaic effect is a process where electrons absorb photons (bundles of light energy), exciting them to a higher energy level. The absorption allows the electrons to move more easily across the junction between the two materials. This process generates a voltage and a direct current.

Photovoltaic technology has been around for a long time–almost 160 years. Knowledge of how certain materials produce an electric current when exposed to light was first discovered in 1839 by a 19-year-old French physicist, Edmond Becquerel. The theory behind it was further refined in 1905 by Albert Einstein, who published a paper on the nature of light and the photoelectric effect.

About 45 years later, Bell Laboratories was experimenting on silicon semiconductors when it discovered that silicon could convert light to electricity. It quickly established the first silicon PV cell in 1954.

The first silicon PV cells were expensive and had just a 4% efficiency in converting sunlight into electricity; however, this changed during the 1960s space race, when the National Aeronautics and Space Administration (NASA) utilized them to power space satellites.

NASA's interest and research on this technology led to significant improvements, and drove the cost down by 80%. This led to the swift development of solar PV technology for a wide variety of applications. By the 1970s, silicon solar PV technology was deployed for many land-based applications, such as for powering batteries for navigational aids. By the 1980s, further efficiency and other improvements led to its use for small battery charging applications, such as for watches and calculators. By the 1990s, the governments of Germany and Japan, for example, initiated programs for the wide adoption of rooftop solar PV systems.

2.7.2　What is the difference between a solar or photovoltaic cell, module, and panel?

Solar or PV cells are individual semiconductor components in PV modules that convert light energy from the sun into DC electricity through the photovoltaic effect. The cells are connected electrically, assembled in a frame, and sealed from the environment with a protective glass or laminate to produce a module.

A PV panel is an electrically-connected assembly of one or more PV modules mounted on a supporting structure and ready for installation in the field.

2.7.3　What are the types of solar modules, and which one is best?

PV module technologies fall in three categories: first-generation crystalline silicon, second-generation thin film, and third-generation emerging and novel technologies. Descriptions of these different categories and some of the main subcategories follow.

Choosing which one is the best technology for a project depends on a number of factors, including efficiency, price, availability in the market, length and terms of performance guarantee, form and appearance, and response to the climatic conditions. The descriptions that follow compare some of these factors between different technologies, to provide some guidance in choosing which is best for the solar project.

For instance, a module with higher efficiency can generate more power per module area, and thus be useful for rooftops where space is limited. In such a case, comparing efficiencies of different technologies (e.g., crystalline silicon versus thin film) would be helpful. Figure 6 shows a comparison of different technologies' efficiencies, as compiled by the United States National Renewable Energy Laboratory. These data are updated up to the year 2013.

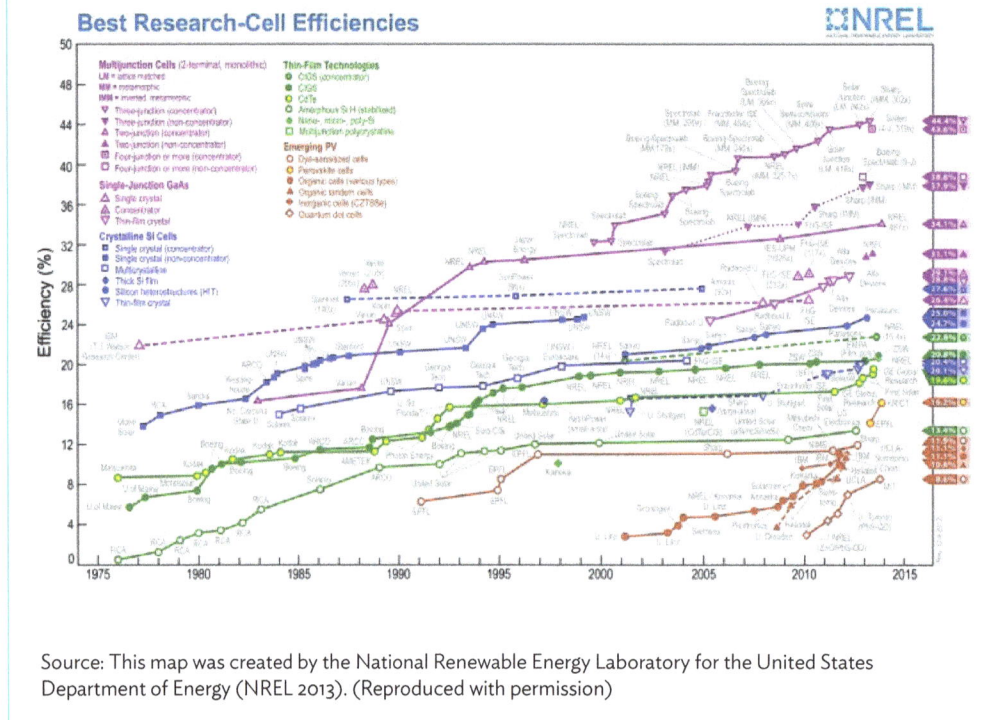

Figure 5: Research Results on Cell Efficiencies over Four Decades, Showing Steady Improvement for Virtually All Photovoltaic Conversion Technologies

Source: This map was created by the National Renewable Energy Laboratory for the United States Department of Energy (NREL 2013). (Reproduced with permission)

Crystalline Silicon
Crystalline silicon cells (i.e., first-generation cells) make use of high-quality yet abundant silicon as a primary material. They tend to be more expensive but also more efficient than thin-film technologies.

The module efficiency of crystalline silicon technologies generally ranges from 14% to 22% (Greenpeace and EPIA 2011). The more efficient technologies are also the most expensive.

The most expensive but the most efficient technology is *monocrystalline silicon.* It is known as the "workhorse" of the industry. It is made from single-crystal wafer cells cut from cylindrical ingots. By comparison, *multi-* or *polycrystalline modules* are less expensive but also less efficient. Their cells are made from square-cast ingots. A third, lower-cost and lower-efficiency technology is *ribbon silicon.* It is made by drawing flat thin films from molten silicon, thus creating a multi-crystalline structure. In all cases, cells are then wired together in a variety of ways to produce a module.

Thin Film
Also known as second-generation cells, thin film technologies are currently the fastest growing sector of the PV industry. They are manufactured by depositing the substrate material onto glass, stainless steel, or plastic. They are lower in cost compared to crystalline silicon, which makes them attractive for utility-scale applications, where the lower efficiencies can be offset by the use of large arrays and where roof space or land availability is not a key issue.

Besides their lower cost, there are a few other advantages. One attribute is their flexibility. The modules are not as rigid as they are with mono- and polycrystalline technologies, and therefore can be "molded" to fit non-flat surfaces. Also, because thin film can be deposited onto lighter substrates such as plastic, they may be a better choice for roofs that are not able to bear the weight of heavier crystalline modules.

Another advantage deals with their suitability for hotter climates. Thin film technology has relatively low temperature coefficients (around 0.2% per degree Celsius. See Table 3), which means that this technology will perform better in hotter climates.

In addition, some thin film modules can be laid directly onto the roof or integrated within the building structure. This is known as building-integrated PV (BIPV). It removes any concern for wind loading.

One important disadvantage to thin film technologies is their reliance on rare earth materials (such as indium and tellurium), which are available in only a few locations. Export for these materials is strictly controlled by the countries where the materials can be found. Experts have predicted that limited availability of rare earth materials could lead to a supply gap, which may cause prices to spike (Quantum Solar Power 2012).

Table 3:   Performance of Photovoltaic Technologies

	Symbol	Temperature Coefficient (%/°C)	Cell Efficiency (%)[a]	Module Efficiency (%)[a]
Crystalline Silicon	**c-Si**	**0.37–0.52[b]**	**N/A**	**N/A**
Monocrystalline	mono-c-Si	~0.45[c]	16–22	13–19
Polycrystalline	poly-c-Si	~0.5[c]	14–18	11–15
Ribbon silicon	N/A	N/A	N/A	N/A
Thin Film				
Amorphous silicon	a-Si	0.1–0.3[b]		4–8
Thin film silicon	TF-Si	0.58–0.62[b]		N/A
Amorphous and micromorph silicon multijunctions	a-Si/μc-Si	0.24–0.29[b]		7–9
Cadmium-telluride	CdTe	0.18–0.36[b]		10–11
Copper-indium-[gallium]-[di]selenide-[di]sulphide	CI[G]S	0.33–0.5[b]		7–12
Emerging and Novel				
Concentrating photovoltaics	CPV	N/A	30–38	~25
Dye-sensitized solar cell (organic)	DSCC	N/A	2–4	

°C = degree Celsius, N/A = not available.

Sources:
[a] Greenpeace and EPIA (2011) (based on various sources).

[b] Virtuani, Pavanello, and Friesen (2010) (based on manufacturer's datasheets and literature).

[c] National VET (2012).

There are four common types of thin film technologies: (i) amorphous silicon, (ii) multijunction, (iii) copper-indium-gallium-selenide, and (iv) cadmium-telluride:

(i) *Amorphous silicon,* or a-Si, consists of a non-crystalline silicon material placed on a variety of surface types. The manufacturing process is highly proven, but currently results in low module efficiencies of around 4%–8%.
(ii) *Multi-junction thin silicon film* is made by depositing a-Si and microcrystalline silicon onto an a-Si cell. The microcrystalline silicon layer increases the efficiency of the cell to 7%–9% because it absorbs additional light in the red and infrared end of the spectrum.
(iii) *Copper-indium-gallium-selenide* (CIGS) is able to achieve the highest efficiencies of thin films of around 7–12%. However, challenging manufacturing issues in maintaining uniformity of the product result in higher costs.
(iv) *Cadmium-telluride* (CdTe) has the lowest manufacturing costs and reasonable efficiencies, ranging from 10% to 11% (Greenpeace and EPIA 2011).

Emerging and Novel Technologies
There are a variety of novel PV technologies (i.e., third-generation cells), emerging on the market. These include concentrating photovoltaics (CPV) and organic solar cells. Novel technologies under development include advanced inorganic thin films and thermophotovoltaic cells.

The technology for third-generation cells tend to be less developed and more expensive compared to first- and second-generation cells. However, they have flexible substrates and perform better in dim or variable lighting conditions.

CPV lenses capture light from a large area and focus direct sunlight on small solar cells. A smaller cell means that more costly and highly efficient cells can be used, resulting in module efficiencies of around 25% (Greenpeace and EPIA 2011). However, because it can only capture direct irradiation, a solar tracking system is necessary and is more suitable in areas with a low fraction of diffuse light than in humid tropical climates.

Organic solar cells can either be fully organic or also contain inorganic materials, such as the hybrid dye-sensitized solar cell (DSSC). Organic materials are cheaper to procure, and the manufacturing process has the potential to be low-cost. However, the efficiency is only around 2%–4% for DSSC (Greenpeace and EPIA 2011). They are also instable over time.

2.8   Mounting System Design

The roof type is the major factor for selecting a mounting system. Considerations include whether the roof is flat or pitched at an angle; whether or not the roof can be penetrated; roofing materials; and locations of structural support. Luckily, prefabricated mounting systems come in many forms and can accommodate these structural variations.

The designer of the mounting system should coordinate closely with the roofer, particularly if the roof is still under installation warranty or manufacturing warranty, as the installation of a solar system could void the warranties. Often, the warranties will specify that a certified roofer carry out the installation or seal the roof.

Flat roofs. For flat roofs, such as the roof at ADB headquarters (Box 11), a main issue is whether the roof can be penetrated to fasten the mounting system. A mounting system that can penetrate the roof and be physically attached to the roof or its underlying structure will be lighter and more secure. While one could utilize ballasted mounting frames that do not penetrate the roof, the ballast (typically made of concrete) adds additional deadweight that the roof must withstand.

To reduce the weight of a ballast system, an interlocked mounting system to reduce the amount of ballast can be utilized. Also, a mounting method called "minimally attached" can also reduce weight. It uses both ballast and structural attachments.

Box 11: Mounting System for the ADB Rooftop Solar System

The Asian Development Bank (ADB) installed its photovoltaic system on the roof of the ADB Facilities Block. The roof would be considered flat for installation purposes, although it is slightly curved to allow for water drainage.

Penetrating the roof was a major issue. The roof contains a waterproof membrane underneath concrete tiles. Because penetration of the roof would put holes in the waterproof membrane and could result in a leaking roof, the module mounting frames were fixed to the roof by ballasting with concrete blocks.

The ballast method is less secure, and Manila is at risk of experiencing major seismic events. Thus, using a ballast method for mounting would result in a small risk that the system would slide along the roof, which could damage or disable the array. Rubber feet were added, to increase the friction between the frame and the concrete roof tiles.

These deflectors were tested in 2014 when Typhoon Rammasun (locally known as Glenda), a category 4 supertyphoon, struck Manila. The 200 km/h winds brought by the storm caused extensive damage to Manila, collapsing power lines to the point that 90% of the city lost electricity. Despite the strength of the storm, the rooftop solar panels were undamaged.

Source: ADB.

Pitched roofs. Mounting methods for pitched roofs can be classified into two broad categories: (i) integral mounting, where the mounting is integrated and flush with the roofing or exterior of the building itself; and (ii) standoff mounting, where the modules are located at least three inches above the roof surface.

Building owners may prefer integral mounting for its aesthetic appeal, but standoff mounting will perform better particularly in hot climates, since it allows air to flow around the panels and reduce their temperature. Often, modules are installed parallel to the roof surface. It is simpler, and alternatively tilting the panels does not generally increase the amount of collected energy enough to make it worthwhile.

Roof penetrations. For installations with roof penetrations, a structural engineer can help to design a system that can withstand the dead load and wind loads, and that has fewer but stronger attachments to minimize the risk of leaks. The roof should undergo leak testing after the mounting system has been connected to the roof and sealed, but before the modules are installed.

Building-integrated photovoltaics. In building-integrated PV (BIPV) systems, the modules replace roofing materials or windows, and can even be partially transparent, allowing a portion of the light to pass through the PV panel. Some thin film modules also allow for attachment directly to the roofing surface, removing the need for a mounting frame.

2.9 Inverter Selection

A solar inverter operates much like the components used in a modern large television set for power supply (that convert AC to DC) and amplifiers (that amplify and modulate signals). A solar inverter, technically called a power conditioning unit, converts DC electricity from the solar PV array to AC electricity, while maximizing plant output. It comprises solid-state components, the main one being a bank of thyristors.

Thyristors are switching devices for converting the current flowing along two wires in a DC system to three wires in an AC system. Logic circuits (like computers) control the thyristor gates, as well as the inverter output.

Both thyristor gates and inverter output require minimal attention once their performance has been adjusted to meet the requirements.

2.9.1 Types of Inverters

Three types of solar inverters are in common use: string, central, and micro inverters. They can be differentiated by their wiring configuration:

- a *micro inverter* converts electricity from a single panel;
- a *string inverter* converts electricity from a single string of modules (a string is a set of modules wired in series); and
- a *central inverter* converts electricity from multiple strings wired in parallel to each other (Figure 6).

The ADB Headquarters Rooftop Solar Project has string inverters.

Micro inverters are installed near each panel or attached to the back of the panel. Although more expensive than central or string inverters and more intensive to install and maintain, they perform better than central inverters, especially in shaded conditions—performance is higher by over 20% in unshaded conditions and over 27% in shaded conditions (Lee 2011). In addition, in the case of a fault with one panel, the power from the other panels can still deliver to the load.

Central inverters are the most common, particularly for medium- and large-scale plants. Installation is simpler for central inverters than for string and micro inverters, but central inverters need a longer lead time for repair because specialists are required. In contrast, string inverters can be quickly replaced by the project developer and/or maintenance crew.

In choosing between string or micro inverters and central inverters, the former are likely to be a better option for roofs that are partially shaded during parts of the day, or for arrays that contain modules with different specifications or orientations. This is because the inverters are able to track and maximize power output for each module or string, rather than for the array as a whole (IFC 2012).

In addition, a solar PV system will require a *grid-tie inverter* if the system is grid-connected. A grid-tie inverter synchronizes the AC phase from the solar power plant with the utility phase, and shuts down automatically when the grid is not supplying electricity. This "anti-islanding protection" protects electrical workers who repair power lines. It also protects equipment and appliances from a fluctuating power supply.

A grid-tie inverter should be compliant with the following:

(i) the UL 1741 Standard for Safety of Inverters, Converters, Controllers and Interconnection System Equipment for Use with Distributed Energy Resources as published by Underwriters Laboratories Inc. on 28 January 2010; and
(ii) the IEEE 1547 Standard for Interconnecting Distributed Resources with Electric Power Systems.

Although these are standards in the United States, other countries accept them. Even if a country does not have equivalent national standards, an inverter that conforms to one of the above (or similar) standards ensures protection of both the utility grid and the local building grid.

Figure 6: Solar Inverter Configurations

(a) String Inverters

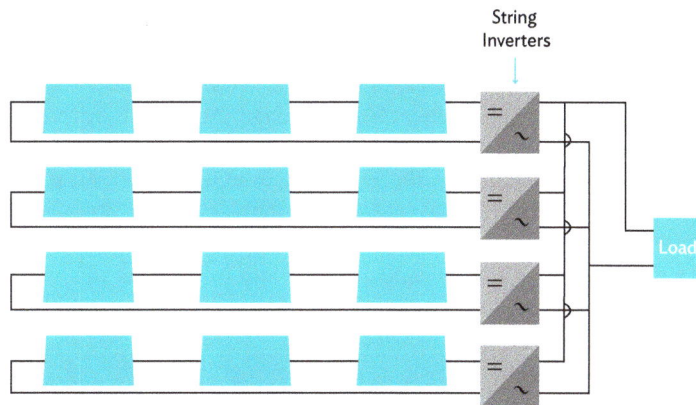

(b) Central Inverter

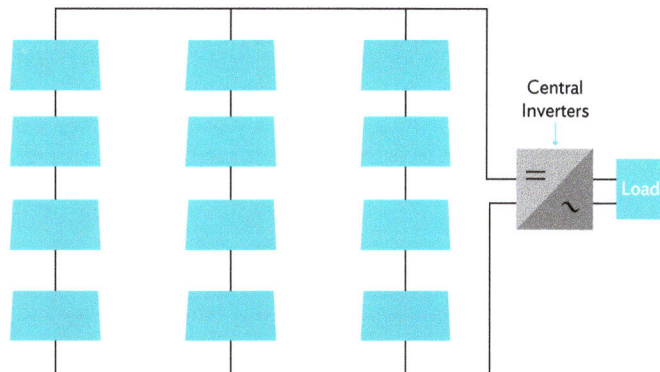

(c) Micro Inverters

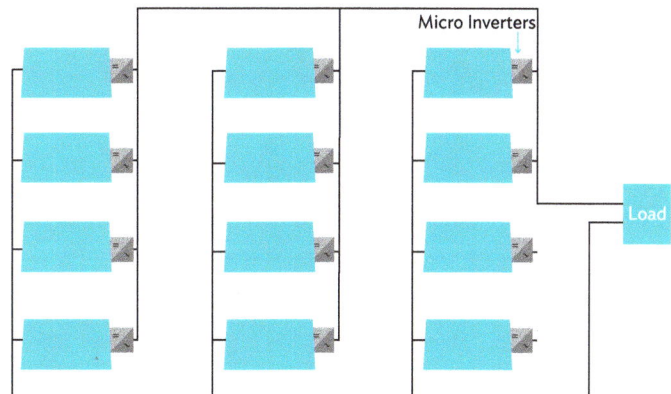

Source: ADB.

Specific requirements for inverters have to be obtained from the national grid, electrical codes and the electrical utility. For instance, some utilities require inverters to be procured from their list of approved models.

2.9.2 Inverter Selection: Inverter Capacity and Number of Inverters

With regard to capacity, a grid-tie inverter should have the capacity to convert the full output of the solar PV array and have the quality of utility-grade electricity.

As to number of inverters, small solar PV installations of 1 kilowatt capacity or less will only require a single inverter, while larger installations will require a number of inverters due to the following constraints:

(i) **Reliability.** Multiple inverters are more reliable than a single inverter system. If the single inverter fails, the whole system goes down. If one inverter out of many fails, then only the power directed through that inverter is lost.

(ii) **Efficiency.** Inverters with a large capacity are more efficient than smaller inverters. The size of the inverter must be matched with the array that it will handle.

(iii) **Solar PV array segmentation and shading characteristics.** Shading of part of a string of PV modules in an array reduces the output of the whole string. Segmentation of the PV array will isolate the shaded modules from the rest of the PV array, reducing the effects of shading. The number of inverters can be matched with the number of array groupings based on the shading characteristics of the array. The capacity of the inverter should be matched to the size of the sub-array that it is assigned to handle.

(iv) **Space constraints.** Fewer larger inverters will occupy less total space than several small ones. However, with limited space on rooftops, a big enough space may not be available for larger units. Large inverters will also require special equipment for installation—such as forklifts, cranes, or chain blocks—that the available space or structure may not be able to accommodate.

(v) **Environmental conditions.** Inverters are ideally housed in power control rooms with the rest of the switch gears of the facility. Outdoor units may be practical if indoor space is limited. Extra protective coating for the inverters is required in corrosive environments.

2.9.3 Input Voltage and Capacity Considerations

Which inverters are right for the system depends on the PV array configuration and output. Considerations include the operating voltage, current, and power output of the array, which should always be within the operating range of the inverters.

Often, inverter manufacturers provide string sizing guidelines or online programs that perform the calculations discussed below.

Maximum Voltage
Calculating maximum voltage considers that the output voltage from a solar module increases as the temperature decreases at a rate defined by the temperature coefficient. The historical extreme minimum air temperature T_{min} (in degrees Celsius) for the location[14] may be used to calculate the maximum number of modules that can be installed while remaining below the maximum rated voltage of the inverter.

Grid-connected inverters are typically designed to handle up to 600 volts open circuit (V_{oc}).

---

[14]   The historical minimum and maximum ambient air temperatures may be obtained from the national meteorological organization for the weather station nearest to the site.

For a single solar module, the adjusted voltage $V_{adj,max}$ at the minimum temperature is:

$$V_{adj,max} = V_{oc} + (T_{STC} - T_{min}) \times T_k V_{oc}$$

where:

- T_{STC} is the standard test condition temperature in degrees Celsius (typically 25°C);
- V_{oc} is the open circuit voltage of the module at standard test conditions in volts; and
- $T_k V_{oc}$ is the temperature coefficient in volts per degree Celsius,[15] all of which should be specified on the module datasheet (Mayfield 2008).

The maximum number of modules $N_{s,max}$ that can be linked in series is then:

$$N_{s,max} \leq \frac{V_{max}}{V_{adj,max}}$$

where V_{max} is the maximum voltage the inverter can support. Rounding $N_{s,max}$ down to the nearest whole number ensures the maximum voltage will not be exceeded.

Minimum Voltage
The minimum array operating voltage must be higher than both the minimum operating voltage and the inverter start-up voltage (which could be higher than the minimum operating voltage) to enable energy to be supplied the whole time the array is generating electricity. The PV array will have a lower operating voltage when the cell temperatures are high, and also when solar radiation levels are low especially during the early morning, late afternoon, and on cloudy days.

The adjusted voltage $V_{adj,min}$ for a single module at the minimum temperature is:

$$V_{adj,min} = V_{oc} + [T_{STC} - (T_{max} + T_{cor})] \times T_k V_{oc}$$

where:

- T_{max} is the historical maximum ambient air temperature; and
- T_{cor} is a correction for the cell temperature. It is 35°C if the modules are mounted parallel to the roof with less than 6 inches (15 centimeters) of airflow space in between the panels and the roof, and 30°C if the modules are further from this on a rack-type mount (Mayfield 2008).

Calculate the minimum number of modules $N_{s,min}$ using

$$N_{s,min} \geq \frac{V_{min}}{V_{adj,min}}$$

Round $N_{s,min}$ up to the nearest whole number to ensure the system always operates above the minimum voltage.

Since modules degrade over time by up to 1% of the voltage per year, select the number of modules in each string so that it will operate well over the minimum voltage (Mayfield 2008).

---

[15] If the temperature coefficient is given in percent per degree Celsius, convert to voltage per degree Celsius by multiplying by the open circuit voltage.

Maximum Current

The inverter can withstand a maximum input current I_{max}, which can be controlled by limiting the number of strings in parallel:

$$N_{p,max} \leq \frac{I_{max}}{I_{module}}$$

In this equation, the module current I_{module} is either the rated or short circuit current depending on what is specified in the inverter manual (Berdner 2009).

Maximum Power

For the full delivery of power, it is important for the inverter to be able to handle the power input. If the power input to the inverter exceeds the inverter's output power capability, the excess energy will be dissipated or lost as it passes through the inverter.

Output Parameters

The electricity output from the inverter must match the interconnection requirements of the grid and of the internal power system. The inverter manufacturer will often set up the inverter to match these requirements prior to shipping and installation, and will also participate in the commissioning process to ensure proper configuration and use.

The on-site performance of the power inverter should be certified by an international independent third party certification body, e.g., Fraunhofer Institute of Germany, TÜV of Germany, Solar Institute of Singapore, Arsenal of Austria, or Underwriters Laboratories (UL).

2.10   Wiring Design

There are a number of considerations for wiring design:

- The system designer must be familiar with all of the parts of the building and electrical codes pertaining to solar, and the electrical wiring must comply with these codes.
- The PV module arrays have to match the properties of the inverter (described earlier).
- The electrical wires and cables connecting the inverters to the building switch gear have to be sized to minimize losses and ensure electrical safety.
- The wires need to be rated to withstand the high ultraviolet radiation and heat they will be exposed to on the roof (Gevorkian 2008).

Another consideration is grounding. More care needs to be taken in grounding a rooftop system than a ground-mounted system because it is more insulated from the ground. Unless specified otherwise in the electrical code, grounding wires from all of the equipment need to be connected together and directed to a single service grounding point (Gevorkian 2008).

Lightning affecting the equipment will also affect wiring. Lightning subjects equipment to high voltages and currents, which can break down insulation between circuit elements and cause serious damage to the system components. Therefore, protection against lightning is necessary, especially in tropical areas like Manila.

Surge protectors or lightning arresters and special grounding will protect equipment from damage by lightning. They work by providing a direct conduction path to the ground, so that the solar PV system components would not be struck by lightning (Gevorkian 2008).

2.11   System Performance Assessment

The energy produced by a PV system can be estimated from the rated power of the installed system, the solar resource, and the derating factors. This estimate must use the specific selected system design and individual losses. It is recommended that conservative estimates of system performance or a range of expected output values be used. Many available computational tools can assist in this estimation.[16]

2.11.1   Rated Power of the Installed System

PV modules are rated under specific standard test conditions (STC) that are easily recreated in a factory and allow for consistent comparison of products. Since PV modules produce DC electricity, their rating is designated in terms of watts DC at STC.

To determine the total DC power rating of a PV array, multiply the number of panels in the array by the DC rating of each panel. An example from the ADB rooftop solar system design is shown in Box 12.

2.11.2   Solar Resource

The energy output from the system is almost directly proportional to the amount of solar resource incident on the solar panels. Solar resource varies from hour to hour, day to day, and season to season.

For smaller systems, the annual GHI should be sufficient to estimate performance of the system, while for larger systems that need more accurate predictions, hourly or daily data may be used.

Section 2.4 outlines the types of data available. The higher-quality and longer-term the data are, the more accurate the prediction will be.

2.11.3   Derate Factors

Derate factors represent the estimated efficiency of a component of, or environmental effect on, a solar PV system. They account for factors that can impact PV system output, including the losses due to shading and module orientation described in Sections 2.5 and 2.6. They therefore should be factored into the system design and energy production estimates.

The optimal derate factor is 1.00. A value less than 1.00 represents reduced system output power. For example, if the power lost through a particular component is 2%, then the derate factor for that component is 0.98.

Obtaining the derate factor for individual components, as well as for the whole PV system, is helpful. Obtaining the derate factor for the whole system accounts for all losses between the DC system nameplate power and the AC output power. It can be calculated by multiplying all of the individual derate factors.

Shading. The derate factors for shading are the most crucial to include in system design and energy production estimates. This is because shading from nearby buildings and vegetation, obstructions on the roof, and adjacent rows could cause significant losses.

The roof cover ratio, which is the ratio of the PV area array to the total area of the rooftop,[17] is a useful indicator for estimating losses due to shading from adjacent rows. Standard industry practice is to target a loss of 2.5% (derate factor of 0.975) by optimizing roof space.[18] To limit loss to this value, the roof cover ratio should be at least 0.7 for an array with a fixed tilt of 10° to the equator, 0.55 for a 20° tilt, and 0.45 for a 30° tilt.[19]

---

[16] Commonly used free tools are PV Watts (http://www.nrel.gov/rredc/pvwatts) and the Systems Advisor Model (SAM, https://sam.nrel .gov). Many free tools are also available such as PVSyst (http://www.pvsyst.com/en).

[17] An analogous "ground cover ratio" is commonly used for ground-mounted systems.

[18] See http://rredc.nrel.gov/solar/calculators/tts/system.html

[19] See http://rredc.nrel.gov/solar/calculators/pvwatts/system.html

Box 12:   Combined Rated Power of the Solar Panels Used for the ADB Rooftop Solar System

The Asian Development Bank (ADB) system has 2,040 modules each rated at 280 peak watts, combining to make a total of 571 kilowatts-peak. The module area is 1,946 millimeters (mm) x 992 mm = 1.94 square meters (m²), so the conversion efficiency of each module at standard test conditions (STC) (1,000 watts per square meter) is 14.4%.

The modules are mounted on a galvanized steel frame and concrete ballast structure, and cover 5,600 m² of roof area. The panels themselves cover 70% of this roof area.

The system uses 34 inverters to convert direct current (DC) electricity generated by the photovoltaic (PV) arrays into alternating current (AC) electricity that is connected to ADB's existing power system. The expected optimal performance ratio was 74.5%,[a] equivalent to an expected output power of 425 kilowatts.

Specifications of the Module Used in the ADB Rooftop Solar System

Item	Units	Value
STC power rating (P_N)	Wp	280
Peak efficiency	%	14.4
Power tolerance	±%	+5
Number of cells	No.	72
Temperature coefficient of P_{max}	%/°C	−0.45
Temperature coefficient of V_{oc}	%/°C	0.33
Temperature coefficient of I_{sc}	%/°C	0.055
Current at maximum power point (I_{mpp})	A	7.95
Voltage at maximum power point (V_{mpp})	V	35.2
Short circuit current (I_{sc})	A	8.33
Open circuit voltage (V_{oc})	V	44.8
Maximum system voltage (V_{max})	V	1,000

°C = degree Celsius, A = ampere, V = volt, Wp = peak watt.

Note: The expected performance ratio was calculated using a simulation that described optimal design conditions for the system. In the first year of operation, the system operated below this expected performance, generating a net AC output of 376 kilowatts, equating to a performance ratio of 65.8%. Note that the actual weather measurements were not used in the simulation and thus may be a factor in the lower performance ratio.

Source: ADB.

Reflection. In calculating sunlight reflection, the reflection losses due to diffuse and direct irradiation should be calculated separately. Reflection losses from direct irradiation on standard module types can be calculated using the known reflection as a function of the angle of incidence for a low iron module glazing. Constant reflection losses of 3.5% can typically be assumed for the direct and diffuse irradiance combined.

Temperature. Derate factors for temperature take into account output power decreases, as module temperature increases. The temperature coefficient, listed on the module datasheet, is the measure of the amount of loss for every degree that the module is above 25°C (Box 13). Typical temperature coefficients are −0.45%/°C for monocrystalline modules, −0.5%/°C for polycrystalline modules, and −0.2%/°C for amorphous modules (Table 3).

Box 13:   Power Output and Temperature for the ADB Rooftop Solar System

For the Asian Development Bank (ADB) rooftop solar system, the actual direct current power generated is less than the power rating of the photovoltaic (PV) modules. However, the PV modules generate power above their nameplate rating when the ambient temperature is cooler than standard test conditions, which has been observed during the cooler months between October and February.

Source: ADB.

Soiling. This accounts for dirt, dust, bird droppings, and other materials that have fallen on the module, which block sunlight and reduce power output. Inclination angles of at least 10° allow "self-cleaning" during rainfall. Even so, practical experience from other installations show that a 2% annual loss (derate factor of 0.98) should be assumed from soiling.

Mismatch and wiring. Adjacent modules do not perform identically, and this mismatch typically results in at least a 2% loss (derate factor of 0.98). Additional losses exist due to resistance in system wiring. Generally, a larger wire will have greater conductivity and therefore lower losses. However, wiring losses may be minimized by properly sizing the wiring for the system.

Inverter. Inherent losses exist in converting the DC electricity generated by the PV modules to the required AC electricity through an inverter. Most inverters have peak efficiencies of over 90%.

Module degradation. This accounts for power output decreases over time due to wear and tear on the modules. The amount of degradation depends on the model, a reduced performance of about 1% per year can be expected. Most module manufacturers will provide a guarantee for 90% of minimum peak power in the first 10 years, and 80% for the next 10–15 years.

2.11.4   Estimation of Alternating Current Energy Output

To estimate energy output from the system over a given time period, multiply the rated power of the system C_R (kWp)[20] by the total GHI_a for the location over that period (kWh/m²) and by the DC to AC system derate factor D:

$$E = C_R \times GHI_a \times D$$

For instance, a 1 kW rated system operating at a site with an average solar resource of 5 kWh/m² per day would ideally produce on average an energy output of 5 kWh per day. If the derate factors are taken into account, the actual energy production would be 5 kWh/m² per day multiplied by the derate factors. If all the losses associated with shading, electrical losses, etc. total 10%, the derate factor would be 0.9 and the actual output of the system would be 4.5 kWh per day.

Box 14 shows another example of how AC energy output can be calculated, using the case of ADB's rooftop solar.

[20] The rated power of the system is equal to the STC irradiation (1 kW/m²) multiplied by the array area (in square meters) and the panel conversion efficiency.

Box 14:   Derate Factor for the ADB Rooftop Solar

The estimated direct current (DC) to alternating current (AC) derate factor from the third party verification prior to finalization of design was 0.684, due to the following factors:

- external shading (10.2%),
- internal shading (3.6%),
- soiling (2.0%),
- reflection (3.6%),
- spectral losses (1.0%),
- irradiation-dependent losses (3.9%),
- temperature (7.8%),
- mismatch (0.8%),
- cabling (0.8%), and
- inverter (2.8%).

The Asian Development Bank's (ADB) rooftop solar has a rated power of 571 kilowatts-peak (kWp), so these losses would result in an actual output of 390 kW.

In the first year of operation, the total incident solar radiation measured by a pyranometer at the site was 1,675 kilowatt-hours (kWh) per square meter, and the total energy output measured at the net meter was 630,000 kWh.

The DC to AC system derate factor, D, was thus:

$$D = \frac{630{,}000}{571 \times 1{,}675} = 0.66$$

Source: ADB.

2.12   Due Diligence

Due diligence is the independent evaluation of the expected performance of a project. Conducting a due diligence evaluation is required for seeking financing or for a project that will enter a power purchase or solar lease agreement.

The evaluation needs to contain an unbiased opinion of the expected technical performance and long-term cash flow. Critical components of a report would include the following:

(i)   evaluation of meteorological data (especially solar and temperature data);
(ii)   evaluation of potential system losses;
(iii)   estimates of system performance; and
(iv)   estimates of the project cash flow over the projected life for the project, with particular attention given to periods when cash flows could be negative due to financing cycles or low solar resource periods.

As such, an independent consultant can be hired to carry out the evaluation, provided the consultant has knowledge of resource assessment and system design tools.

Typically, financiers will require a due diligence evaluation be conducted after the project receives internal approval and the system has been designed. ADB hired an independent party to conduct a due diligence evaluation following bidder selection (Box 15).

Once the system design has passed performance evaluation, it should also undergo an independent structural and seismic analysis.

> ## Box 15:   Independent Evaluation of the ADB Rooftop System Design
>
> An independent party evaluated the selected bidder's design prior to finalization of system design and signing of the contract. The evaluation included, but was not restricted to, the following areas:
>
> (i)   solar irradiation databases and assumptions;
> (ii)   material and system losses, including shading analysis;
> (iii) electromechanical design as adapted to site conditions:
>   (a) sizing of solar photovoltaic modules;
>   (b) sizing of balance of system;
>   (c) wiring and/or cabling; and
>   (d) support structures.
> (iv)   annual electrical yield output (kWh/year); and
> (v)   techno-financial model.
>
> This report provided valuable input to the project, since it produced a slightly lower estimate of the expected system performance. It also provided some suggestions for improving the design.
>
> After the selected bidder finalized the installation design, the Asian Development Bank forwarded it to an engineering company for conducting a peer review of the structural design. It confirmed that the system and ballast could withstand expected wind loadings. However, it helpfully recommended monitoring of the deflection of roof trusses, due to the additional weight of the system. It also importantly found that the system would not be secure in a seismic event, but that movement and resulting damage to the system would not pose a significant risk to human life.
>
> Source: ADB.

Chapter 3: Procurement

Procurement refers to the process of obtaining the contractors needed to implement the project. Procurement can be anticipated for purchasing photovoltaic (PV) modules and balance of system, the design and construction process, and the operation and maintenance phase.

Procurement can take place separately for each stage of development, or be conducted under a single process. The Asian Development Bank's (ADB) rooftop solar PV project chose a single procurement process, since that would streamline project development. In ADB's case, the project developer selected PV panels and balance of system, designed and implemented the project, and operated and maintained the system.

ADB's procurement process is described in Box 16. That process utilized international competitive bidding, with a one-stage and two-envelope bidding procedure. The latter refers to the technical and the commercial bids being submitted at the same time but in different envelopes.

Box 16: Procurement Process for the ADB Rooftop Solar Power Project

The whole procurement process took 10 months, from designating a procurement committee through contract signing.

Designating a procurement committee involved assigning procurement responsibilities to the Institutional Procurement Committee. Its purpose is to manage the Asian Development Bank's (ADB) procurement of goods and services for its own institutional use. This committee operates under the Office of Administrative Services in ADB, and consists of eight ADB staff members from different departments. The members included a legal adviser and technical staff.

The committee's procurement process involved the following: (i) preparation of procurement procedures, (ii) preparation of the bidding document (available upon request from ADB), (iii) internal approval to move forward with the bidding process, and (iv) and publication of an Invitation to Bid on ADB's website. A sample Invitation to Bid is available in Annex 6.

ADB charged a nonrefundable fee to interested parties wanting to obtain the bidding documents. The fee helped to identify serious bidders that had the capacity to implement the project.

After a month of issuing the Invitation to Bid and allowing potential bidders to obtain the documents, ADB held a question and answer session for them. It allowed ADB to clarify administrative and technical requirements, and other information that would assist bidders in bid preparation. After the prebidding session, ADB distributed the queries and responses in written form to all parties who had obtained the bidding documents. These parties included those that could not attend the meeting.

Source: ADB.

For smaller projects, it is possible to directly source the solar PV system from a single contractor without undergoing the bidding process. However, for larger projects like ADB's Rooftop Solar Project, a bidding process provides more competitive options.

The bidding process typically has the following timeline:

(i)   nominate a procurement committee;
(ii)   publish advance procurement notice;
(iii)   prepare bidding documents and the bid evaluation criteria;
(iv)   publish and issue invitation to bid and bidding documents;
(v)   hold prebid meeting;
(vi)   collect queries and requests for clarification, and disseminate responses to bidders;
(vii)   receive bids;
(viii) open and evaluate the technical bids, and clarify as needed;
(ix)   rank the technical bids and then open the commercial bid, and prepare the final ranking;
(x)   select bidder;
(xi)   negotiate contract;
(xii)   conduct independent evaluation;
(xiii) revise design; and
(xiv)   sign contract.

3.1   Preparation of Bidding Documents

Well-prepared bidding documents help streamline the procurement process, thus saving time and effort for everyone involved.

Being well prepared with the bidding documents first necessitates being familiar with the types of documents required. Table 4 provides a list and description of the essential bidding documents. These are (i) the instructions to bidders, (ii) the bid datasheet, (iii) evaluation and qualification criteria, (iv) bidding forms, (v) a technical outline, and (vi) a draft contract.

Table 4:   Bidding Documents

Document	Description
Instructions to bidders	Specifies the course of actions to be taken by bidders in the preparation and submission of their bids. Also provides information on the submission, opening, and evaluation of bids and on the award of contract.
Bid datasheet	Consists of provisions that are specific to the procurement and supplements the information or requirements in the instructions to bidders.
Evaluation and qualification criteria	Contains all the criteria that will be used to evaluate bids and qualify bidders.
Bidding forms	Contains forms that are to be completed by the bidder and submitted as part of the bid.
Technical outline	Provides basic information to the bidders regarding design processes, installation, testing and commissioning, and operation and maintenance of the solar project.
Draft contract	Contains the terms and conditions of the agreement proposed between the building owner and the successful bidder (power provider) for the lease of the roof space, equipment installation, and project operation.

Source: ADB.

The level of detail in the bidding documents will depend on the project and on individual institutional requirements. However, the technical outline should include all of the information and requirements for the rooftop solar system collected when designing the system (see Chapter 2). This includes any information on required licenses, relevant sections from electrical codes, renewable energy policy and regulation, and meteorological data. Box 17 describes the technical outline that ADB produced.

Box 17: Technical Outline for the ADB Rooftop Solar Power Project

The Asian Development Bank (ADB) developed a project technical outline to provide basic information to bidders regarding ADB's roof, required technical specifications, design processes, installation, testing, and operation and maintenance of the solar project (included in Annex 5). In this document, ADB also included additional relevant information that it had collected on permits, codes, renewable energy policy and regulation, and solar resource data.

This technical outline was included in the bidding documents for providing sufficient information and guidance to bidders for designing a suitable system. This method allowed ADB to clearly set project requirements, while utilizing the bidders' expertise in designing solar power systems. Further, providing the technical outline assisted international bidders with understanding the conditions for implementing a rooftop solar photovoltaic project in the Philippines.

Source: ADB.

3.2 Pre-bid Activities

As demonstrated through ADB's experience (Box 16), holding a question and answer session for interested bidders can prove very useful. As discussed earlier, the aim of this session is to answer questions, emphasize the administrative and technical requirements of the bid, and clarify bid preparation requirements.

To ensure fairness and transparency, all communications should be provided to interested parties in writing and/or uploaded on a website. This should be done within a reasonable amount of time and giving bidders enough time to review the information prior to submitting bids.

Prospective bidders should also have the opportunity to visit and visually inspect the roof.

3.3 Bid Evaluation

There are a number of bidding methods to choose from. These include single-stage, one-envelope; single-stage, two-envelope; two-stage, two-envelope; and two-stage bidding procedures. Definitions of these bidding procedures can be found on ADB's website.[21]

In the case of ADB, it used the single-stage, two-envelope bidding procedure (Figure 7). In this procedure, bidders submit two sealed envelopes simultaneously. One contains the technical proposal and the other contains the price proposal (in this case, the rate of electricity to be sold to ADB, expressed as peso per kilowatt-hour), enclosed together in an outer single envelope.

---

[21] ADB. Bidding Procedures. http://www.adb.org/site/business-opportunities/operational-procurement/goods-services/bidding-procedures

Initially, only the technical proposals are opened at the date and time advised in the bidding document. The price proposals remain sealed and are held in custody by the purchaser (building owner). The purchaser initially evaluates only the technical proposals and rejects deficient proposals that do not meet technical requirements. Unlike two-stage bidding procedures, no amendments or changes to the technical proposals are permitted.

The objective of the exercise is to allow the purchaser to evaluate the technical proposals without reference to price. Bids rejected for not conforming to the specified technical requirements have the price proposal returned to the bidder unopened.

Following the purchaser's approval of the technical evaluation, and at a date and time advised by the purchaser, the price proposals are opened in public. The purchaser then evaluates the price proposals, and awards the contract to the bidder with the lowest evaluated substantially responsive bid.[22]

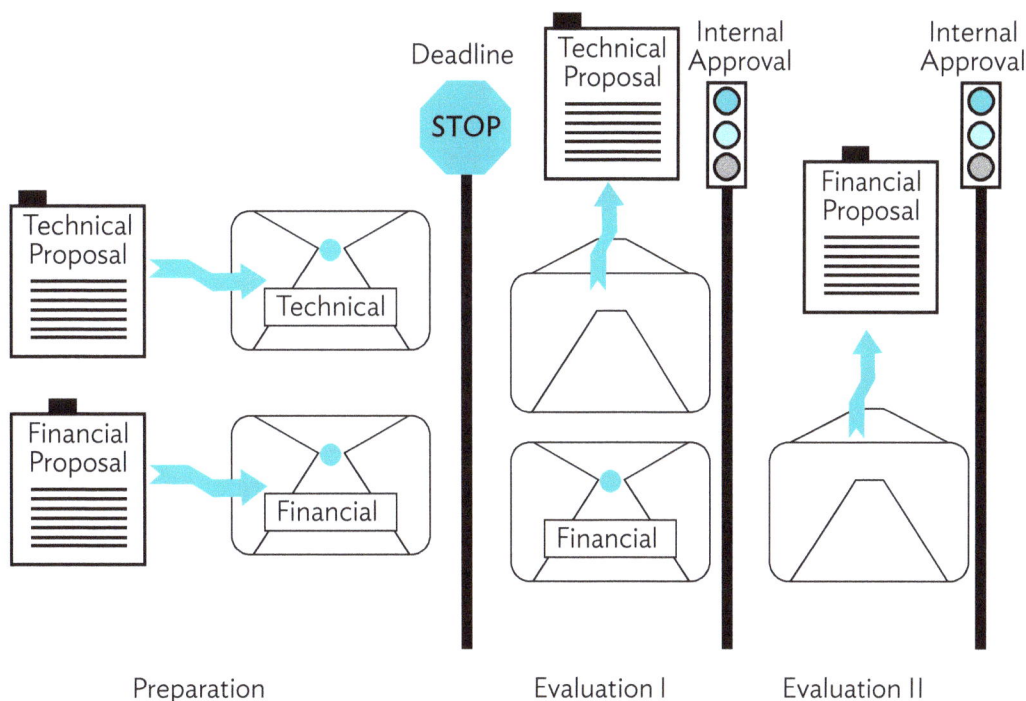

Figure 7: Single-Stage, Two-Envelope Bidding Process

Source: ADB.

---

22 ADB. Bidding Procedures. http://www.adb.org/site/business-opportunities/operational-procurement/goods-services/bidding-procedures

Below are the steps for evaluating the technical proposal and the price proposal under a single-stage, two-envelope bidding procedure.

Technical Proposal Evaluation Process
1. **Open and examine technical bids for completeness.** Open and examine each technical bid to confirm that all documents requested in the bidding document have been provided and are complete. If any of these documents or information is missing, the bid is deemed nonresponsive and the bid may be rejected.
2. **Evaluate technical aspects of bid.** Conduct a detailed technical assessment and evaluation of each complete bid to determine if the technical aspects are in compliance with the requirements stated in the bidding document. Bids that do not meet minimum standards of completeness, consistency, or detail are rejected. Where the bid is substantially responsive but clarification is needed on some aspects, the power purchaser can request the bidder to submit necessary information or documentation within a reasonable period of time, and reject bids for which the bidder fails to submit clarifications.
3. **Assess bidder qualification.** Evaluate whether the bidder meets the eligibility and qualifying criteria specified in the bid documents. The bid is rejected if the bidder does not meet the criteria.
4. **Obtain approval.** Get internal approval of the results of the technical bid.
5. **Inform bidders of results.** Send notification to bidders whose bids meet the eligibility and qualifying criteria on the date of opening of financial bids. For all rejected technical bids, return the price bid to the bidder unopened, with reasons for the rejection specified.

Price Proposal Evaluation Process
1. **Open and examine financial bids for completeness.** Open and examine each price bid to confirm that all documents requested in the bidding document have been provided and are complete. If any of these documents or information is missing, the bid is deemed nonresponsive and the bid may be rejected.
2. **Evaluate price bids.** Check financial bids for arithmetical errors, and make any necessary price adjustment due to that and any discounts offered. Convert bids to a unified currency.
3. **Compare bids.** Calculate the final weighted bid score and ranking according to criteria specified in the bidding documents. The method used by ADB is shown in Box 18.
4. **Obtain approval.** Get internal approval of the results of the financial bid.
5. **Inform bidders of results.** Notify the first ranked bidder that its bid was ranked first and of the next steps, such as independent evaluation (due diligence) of the system design, that are to be conducted before proceeding with contract negotiations. Inform other bidders of their ranking.

3.4 Contracting

The essential goal of a contract is to ensure that each party is clear on their contract obligations, and ensures that each party performs their contract obligations by making them legally enforceable. Retaining legal counsel to draft and review contract provisions is strongly suggested. Each jurisdiction has different legal requirements for contracts, and the nuances of what is or is not legally adequate contract language can be very subtle.

Moreover, legal counsel would be able to assist with drafting provisions that would ensure parties perform their contract obligations. This can be the case even if administrative and legal costs are cut by going with a "boilerplate" contract, where most or all of the provisions are laid out in template form (see **Annex 6**). Provisions can always be added, subtracted, or changed from the boilerplate contract to better protect interests and particular needs, and legal counsel can help with that process.

For example, under ADB's rooftop solar project, one issue was whether the contractor would continue to provide adequate operation and maintenance services after the completion of the contract. To protect ADB's interests, the contract provisions contained a *guaranty*. It essentially ensures that payment to the contractor is contingent upon adequate performance of the solar PV system, as specified under the contract terms.

Box 18: ADB Bid Evaluation Process

To evaluate the technical proposals for its rooftop solar PV system, the Asian Development Bank (ADB) first inspected each technical proposal to confirm that all requested documents had been provided and were complete. It then examined each proposal in detail, comparing them against the technical requirements in the bidding document. Each requirement was assessed on a "pass or fail" framework on the information being present and defined conditions being met. ADB sent individual requests to bidders for technical clarification and confirmation on a few requirements, giving the bidders two days to respond.

Of the five submitted bids, three met the technical requirements. After obtaining internal approval of the technical evaluation results, ADB advised the three qualified bidders about the opening date of the financial bids. At this point, ADB also informed the two other bidders that their bids had been rejected for being noncompliant with the bidding documents, and returned their financial bids unopened.

The Institutional Procurement Committee opened the financial proposals in the presence of representatives from each of the three qualifying bidders. ADB ranked the bids based on the capacity (highest kilowatt-hour per year ranked first) and the fixed electricity price over 15 years of operation (lowest dollar per kilowatt-hour ranked first). To optimize use of roof space and maximize performance efficiency, capacity was given a weight of two-thirds and price was given a weight of one-third. In case of a tie, the bid with higher capacity would be ranked higher. The following table illustrates this.

Bidder	Technical Proposal			Financial Proposal		Technical + Financial	
	kW	kWh/year	Points	$/kWh	Points	Total	Rank
A	450	556,527	200	0.42	84	284	1
B	420	519,425	187	0.40	88	274	2
C	400	494,691	178	0.38	92	270	3
D	390	482,323	173	0.36	98	271	4
E	370	459,631	164	0.35	100	264	5

kW = kilowatt, kWh = kilowatt-hour.

After internal approval, the first-ranked bidder paid for an independent third-party techno-economic evaluation; this confirmed the proposed technical design, financial model, and assumptions.

The selected bidder revised and provided more detailed specifications of its design while responding to the points raised in the independent evaluation. After ADB approved the revised design, ADB and the selected bidder negotiated the contract terms and signed the Solar Services and Site Leasing Agreement (see sample in **Annex 6**).

Source: ADB.

Under the terms of the guaranty, the power purchase agreement stipulates that the solar system will produce a certain level of electricity output (in kWh). The contractor guarantees that it will provide at least 90% of that expected output. If the electricity output is lower than the 90% of the expected output, the contractor pays to ADB the difference between $0.379/kWh[23] and the price ADB would have to pay to its other power provider to make up for the difference.

To protect the contractor, the contract allowed for adjustment of the expected output for operating efficiencies being lower than expected. This would include variations in weather, or where the design capacity is overstated. Also, in the event that electricity output is higher than expected, ADB will purchase the excess electricity at the same price of $0.379/kWh.

[23] $0.379/kWh is the price ADB pays to the system owner for electricity generated from the ADB rooftop solar project.

Chapter 4: Implementation

This chapter covers the following activities after finalization of the solar system design until commissioning:

(i)   acquiring equipment,
(ii)   obtaining permits,
(iii) ensuring safety,
(iv)   installing the system,
(v)   testing, and
(vi)   commissioning.

4.1   Equipment Acquisition

The first step in the implementation phase is to secure the various system components.

For imported equipment, be aware of importing rules and regulations, and import taxes. Poor preparation on this aspect could result in a delay at customs and costly storage fees as customs sorts out any issues.

Upon acquiring the equipment, a secure storage space should be made ready for each component, until the time that the equipment can be installed.

4.2   Obtaining Permits

The next step is to obtain necessary permits and clearances. Generally, it is the responsibility of the contractor to obtain permits and clearances, and the building owner should provide the contractor with the necessary documents and information to obtain the permits. A list of permits that were obtained for the ADB Rooftop Solar Power Project is given in Annex 2.

Exactly which permits are needed depend on the jurisdiction in which the rooftop solar system is to be installed. The permits that are required for a specific project, supporting documents for each, processing times, and the order in which they should be applied for should be identified early on in the project development process (refer to "Permits and Licensing" in Chapter 1).

4.3   Ensuring Safety

Installation of a rooftop solar power system has all of the hazards of a regular building site, plus electrical hazards due to the system itself. Prior to installation, all personnel should undergo safety training that will provide a good understanding of basic electrical principles, work site hazards, and general safety precautions and procedures. Although not always legally required, such training is often part of International Organization for Standardization's occupational health and safety management systems. Adhering to its standards can significantly reduce the number of accidents.

There are numerous hazards to protect against. Common nonelectrical hazards include the following (Daystar 1991):

- exposure to hot or cold weather,
- insects that nest or hide in the equipment,
- cuts from sharp edges on the equipment or bumps from falling equipment,
- falls,
- lifting injuries,
- thermal burns from metal heated by the sun, and
- acid burns from batteries.

Much of these nonelectrical hazards can be prevented through awareness and common sense. Simply wearing suitable personal protective equipment (e.g., hard hat, covered shoes, gloves), anchoring equipment at all times, and practicing proper construction site safety procedures will prevent many accidents. Taking precautions from falling off a roof, particularly a sloped roof, could be handled by ensuring all personnel are clipped into a secure point at all times while on the roof.

Taking precautions against electrical hazards while installing a solar system is similar to that of taking precautions against working on any electrical equipment. The hazard for electrical burns and shocks can exist even with just a handful of photovoltaic (PV) modules, while large installations can produce several hundred volts of direct current (DC) and certainly enough current to cause a fatal injury. In addition, victims of a minor electrical shock could incur worse injury as they jump back from the source of the shock. Therefore, a licensed electrician who understands DC systems and the local electrical code should always be present on the site and perform the electrical work.

Gevorkian (2008) gives the following basic safety instructions:

- Do not attempt to service any portion of the PV system unless the electrical operation is fully understood and installers are qualified to do so.
- Use modules for their intended purpose only. Follow all the module manufacturer's instructions. Do not disassemble modules or remove any part installed by the manufacturer.
- Do not attempt to open the diode housing or junction box located on the back side of any factory-wired modules.
- Do not use modules in systems that can exceed 600 volts open circuit.
- Do not connect or disconnect a module unless the array string is open or all the modules in the series string are covered with nontransparent material.
- Do not install during rainy or windy days.
- Do not drop or allow objects to fall on the PV module.
- Do not stand or step on modules.
- Do not work on PV modules when they are wet. Keep in mind that wet modules when cracked or broken can expose maintenance personnel to very high voltages.
- Do not attempt to remove snow or ice from modules.
- Do not direct artificially concentrated sunlight on modules.
- Do not wear jewelry when working on modules.
- Avoid working alone while performing a field inspection or repair.
- Wear suitable eye protection goggles and insulating gloves rated at 1,000 volts.
- Do not touch terminals while modules are exposed to light without wearing electrically insulated gloves.
- Always have a fire extinguisher, a first-aid kit, and a hook or cane available when performing work around energized equipment.
- Do not install modules where flammable gases or vapors are present.

4.4   System Installation

Qualified professionals, including electrical engineers, should install the rooftop solar system, and a construction manager should supervise the daily construction of the plant. The building owner should engage one of their own engineers or a qualified third party to oversee the installation.

Marking and staking. The first thing the contractor will do is to mark and stake out the future location of the frames. This will take longer when the roof is not perfectly flat, as they will have to mark out the deflections and intrusions on the roof before deciding on exact locations. For a new build, this step will likely simply involve marking out the location of the frames per the building plans.

After marking and staking out the roof, the installers will mount the equipment according to installation requirements and procedures from the manufacturer's specifications.

Mounting system. The contractor will install the frame, making any necessary adjustments to ensure that it is level. If penetrating the roof, the installer or a roofing contractor will properly seal any roof penetrations with sealants and sealing methods that have been approved by the roofing industry. The installation should be conducted according to roofing warranty requirements.

Solar panels. The installer will then attach solar panels to the frame.

String wiring. The contractor will connect the panels in strings according to the wiring plans, making sure to properly ground all components. The electrical engineers will test DC electricity output from each panel and each string to ensure proper connection.

Inverters. The inverters should be preset to the grid requirements. Often, the supplier of the inverters will do this prior to shipping and will be present during installation.

Performance monitoring system. The contractor will install any weather monitoring equipment last. This typically consists of a pyranometer (to measure irradiance), anemometer (to measure wind), and temperature sensors for ambient temperature and module temperature. The weather and performance data (inverter, alternating current [AC] switching cabinet, and net metering) will be fed into a web monitoring system. This information will be used to assess system performance.

4.5   Testing and Commissioning

Once all equipment has been installed, a testing and commissioning process will confirm that the installation is safe and complete, ensure safety to equipment and people, and verify that the system performs to design specifications. Each component is visually inspected and tested in turn before the inverters are connected to the grid. Representatives of both the developer and the building owner should attend the testing and commissioning.

The International Electrotechnical Commission (IEC) has produced an international standard (IEC 62446) for system documentation, commissioning tests, and inspection of PV systems (IEC 2009).The system should also be compared against the local electrical and building codes during the testing.

The first step of commissioning is to conduct a general inspection. The inspection will verify that all components have been completely and correctly installed, properly labeled, can withstand weather exposure, and included in the as-built plans, and that the system is aesthetically acceptable.

The next step is to conduct electrical testing of both the DC and AC components. The testing should be conducted on a fine day when there is a good amount of irradiance. Testing should be carried out by an accredited electrical contractor, preferably an objective third party. As during installation, proper safety precautions should be taken, and careful testing that is properly sequenced will prevent damage to people or equipment.

Tests recommended in IEC 62446 include the following:

- **Module string polarity.** Testing the polarity of each string ensures that wires are correctly labeled and connected—if they are connected in reverse, they could damage the inverter.
- **Open-circuit voltage (V_{oc}) and short-circuit current (I_{sc}).** Combined with irradiance and cell temperature, use these to check that each string is operating within manufacturer tolerances.
- **Grounding continuity.** This tests the array to ground resistance, and ensures that adequate grounding has been established.
- **Insulation resistance testing.** This testing determines the condition of the electrical insulation and verifies that no currents are leaking between the conductors and earth. It is also known as "Megger testing" after the mega-ohms that the testing device measures, or "hipot" as it tests for high-voltage potential between two points.
- **Component functionality.** The functions of all switch gear, controls, and inverters, including inverter synchronization and anti-islanding, should be tested.

Once all testing is complete, the system is energized by switching on each of the inverters one by one. The system should be left connected to the grid for observation of automatic disconnection during sunset and reconnection and synchronization during sunrise.

Once testing and commissioning is complete, the developer should deliver as-built plans to the building owner. The exact contents depend on the project, but should at least include the following:

(i) general system data, including rated power, number and type of modules and inverters, contact information, and project dates;
(ii) wiring diagram;
(iii) datasheets for modules, inverters, and any other major components;
(iv) mechanical design of the mounting system;
(v) operation and maintenance documentation; and
(vi) test results and commissioning report.

Box 19 outlines the testing and commissioning process undertaken for the ADB rooftop solar project.

Box 19: Testing and Commissioning of the ADB Rooftop Solar System

The whole testing and commissioning process for the Asian Development Bank (ADB) Rooftop Solar PV Project took 2 weeks. All parties were present during the testing and commissioning, including ADB, the project developer, a locally certified electrical contractor, and the inverter supplier.

The project developer allowed the electrical contractor to conduct visual inspections and testing of the electrical systems: alternating current low-voltage cable, grounding system, molded-case circuit breakers, and main distribution boards. The visual inspections included comparing equipment to drawings and specifications, inspecting physical and mechanical conditions of each component, and ensuring correct connections and labeling. The electrical tests included grounding, continuity, insulation, and high potential test.

After the electrical testing of the system, the inverters were switched on one by one. The system went online on 20 May 2012, and the project developer completed the commissioning process by submitting as-built plans and commissioning reports to ADB.

Source: ADB.

Chapter 5: Operation and Maintenance

Operation and maintenance of a solar system is much simpler than that of traditional power plants. It will typically only include regular monitoring of system performance and occasional cleaning of the solar panels.

5.1 Performance Monitoring

Basic performance monitoring involves extracting data logged in inverters, switches, and meters. More complex systems use data acquisition systems to incorporate measurements from various components and from weather monitoring devices into web-integrated performance monitoring systems. For larger systems, the system operator should monitor performance once a day. If the monitoring system indicates that power production and/or electrical values for particular strings or components are lower than the design range, or if it indicates a specific fault, then, a unscheduled maintenance is needed. The developer should monitor and respond to any such needs in a timely fashion.

In the case of a power purchase agreement, the system operator should submit a report to the building owner once a month summarizing the energy in kilowatt-hours (kWh) delivered to the facilities, upon which payment is based.

To understand performance over a longer period (i.e., monthly or annually), the following key performance indicators can be used: alternating current (AC) kWh production, AC electricity generation effectiveness, and the performance ratio.

Net AC kWh production indicates the overall performance of a system. It represents the net electricity produced by the photovoltaic (PV) system as measured on a net meter. This metric, measured on a daily, monthly, or annual basis, should always be relayed to the building owner. It is particularly important in the case of a power purchase agreement, as it is the basis for billing. The AC kWh production depends mainly on irradiance and the energy conversion efficiency of the solar arrays, inverters, and associated components.

AC electricity generation effectiveness (ACEGE) indicates a PV system's overall effectiveness in converting incident solar resources into AC electricity (Pless et al. 2005). It is calculated as follows:

$$ACEGE = \frac{Net\ AC\ kWH\ Production\ (kWh\ per\ time\ period)}{Total\ Incident\ Solar\ Radiation\ (kWh\ per\ time\ period)} \times 100\%$$

The *performance ratio (PR)* indicates how well the system performs compared with an ideal system in standard test conditions (STC) that has no losses in the balance of system (Pless et al. 2005). It is therefore a good measure of overall losses due to system inefficiencies. It is calculated as follows:

$$PR = \frac{ACEGE\ (\%)}{Rate\ PV\ Module\ Efficiency\ (\%\ at\ STC)} \times 100\%$$

Losses are always unavoidable, so an ideal *PR* of 1.0 is not achievable in practice. Typically, a performance of 80% or more indicates a high *PR*. A *PR* of 60%–80% would be normal, though this may indicate that the system requires maintenance, and anything below 60% would indicate possible problems with the system that would require corrective actions (Honda et al. 2012). The *PR* is analogous to the system derate factor used in predicting expected performance of a system.

Box 20 gives values for the performance indicators achieved during the ADB headquarters rooftop solar system's first year of operation.

Box 20:   Performance of the ADB Rooftop Solar System During its First Year of Operation

The Asian Development Bank (ADB) system's monthly alternating current (AC) kilowatt-hour (kWh) production was highest at 69,317 kWh in April 2013 and lowest at 43,751 kWh in January 2013. Power output was in line with solar irradiance, which means the system performed as expected.

The system achieved AC electricity generation effectiveness (ACEGE) that averaged 9.63%. The system achieved its highest ACEGE at 10.8% in March 2013 and registered its lowest ACEGE at 8.75% in May 2013.

ADB's system had an annual performance ratio (PR) value that averaged 63%. Monthly PR values ranged from a low of 61% in December 2012 to a high of 66% in April 2013. Under optimal conditions, the system design was expected to have an annual PR of 74.5%.

Source: ADB.

5.2   Cleaning

Soiling of PV panels by dust, bird dropping, leaves, or other debris can result in energy losses by as much as 5%–10%. Soiling is especially a problem in dusty and polluted environments. Although rainfall can clean the panels, there is also a possibility that once the rainwater evaporates, there may be residual dirt or salt deposits.

Regular PV array cleaning to remove soiling and residual deposits will prevent losses in annual solar energy production. In most situations, panels can be cleaned with water.

The frequency of cleaning is dictated by monitoring output. However, cleaning times can also be anticipated more or less depending on weather, nearby construction, or other circumstances. For instance, panels will need cleaning less often during the rainy season, as the rain will wash away much of the debris. Cleaning will need to occur more often if there is nearby construction, since it would result in additional dust in the air.

Algae and other plant life that build up on or under the panels may also need removal, particularly in humid climates.

5.3   Diagnostic Testing and Preventive Maintenance

To ensure the long-term reliable operation of the system, preventive maintenance is recommended. It starts with annual, biannual, or quarterly maintenance and diagnostic testing to ensure that all elements of the system are performing properly. Most common problems include the following: screws have worked loose; the load does not operate properly or not at all; the inverter does not operate properly or not at all; or the array has low or no voltage or current.

Personnel conducting the testing and maintenance must follow proper safety procedures, just as during the installation of the system (see Section 4.3).

The general maintenance involves checking the wiring connections to ensure they are still secure, and checking fuses (such as in the surge protectors or the inverters) for any indication of a fault.

The electrical testing includes measurement of DC voltages string by string and ensuring they are within the design specifications. For a system the size of ADB's system, it takes 6 days for six engineers and technicians to complete the diagnostic testing.

If the system is not running to specifications, the parts at fault should be repaired or replaced. Parts should also be replaced as they reach the end of their life span. For example, lightning protection equipment has a life span of 3 years and will need to be replaced.

Conclusion

With energy demand projected to more than double in the Asia and Pacific region by 2035, it is imperative that the region find renewable sources to generate power while reducing greenhouse gas emissions. Continuing on with business as usual is unacceptable, as the harmful effects of climate change have the potential to wipe out the progress this region has worked toward in its fight against poverty.

Sharing practical knowledge on dissemination techniques for renewable energy seeks to meet that urgency. Knowledge allows people to do more, and do it more effectively.

To that end, this handbook seeks to accelerate the adoption of rooftop solar PV systems, through sharing ADB's prescription for how to implement a project. ADB had a positive experience in developing its own rooftop solar power project, and hopes that sharing its firsthand knowledge used in preparing this handbook will lower barriers to implementing this readily-available and clean technology.

LIGHTING THE WAY: ADB'S ROOFTOP SOLAR PROJECT

ADB is "walking the walk" when it comes to clean energy development by supporting the deployment of clean technology throughout developing Asia and the Pacific and by making its own operations green. ADB has instituted energy efficiency and renewable energy solutions at its headquarters in Manila, the Philippines, and the most visible of these are the 2,000 solar panels on its rooftop. These panels provide a portion of the headquarters' power needs and demonstrate the viability of urban solar power at scale.

References

Asian Development Bank. Bidding Procedures. http://www.adb.org/site/business-opportunities/operational-procurement/goods-services/bidding-procedures

Berdner, J. 2009. Array to Inverter Matching: Mastering Manual Design Calculations. *SolarPro*. Issue 2.1. http://solarprofessional.com/articles/design-installation/array-to-inverter-matching (accessed 11 February 2014).

Blacker, N. and C. Hidalgo Lopez. 2011. Assessment of the Inter-Annual Variability of Solar Radiation for Use in Solar Power Plant Energy Yield Predictions in Australia. In Solar 2011, the 49th Australian Solar Energy Society (AuSES) Annual Conference, 30 November–2 December.

California Energy Commission. 2001. *A Guide to Photovoltaic (PV) System Design and Installation*. June. http://www.energy.ca.gov/reports/2001-09-04_500-01-020.PDF (accessed 5 February 2014).

Couture, S., K. Cory, C. Kreycik, and E. Williams. 2010. *A Policymaker's Guide to Feed-in Tariff Policy Design*. National Renewable Energy Laboratory. July. http://www.nrel.gov/docs/fy10osti/44849.pdf (accessed 3 February 2014).

Daystar. 1991. *Working Safely with Photovoltaic Systems*. http://www.wbdg.org/ccb/DOE/TECH/wksafe.pdf (accessed 13 February 2014).

Gevorkian, P. 2008. *Solar Power in Building Design: The Engineer's Complete Design Resource*. New York: McGraw-Hill.

Greenpeace and European Photovoltaic Industry Association (EPIA). 2011. *Solar Generation 6: Solar Photovoltaic Electricity Empowering the World*. http://www.greenpeace.org/international/en/publications/reports/Solar-Generation-6 (accessed 1 February 2014).

Honda, S., A. Lechner, S. Raju, and I. Tolich. 2012. *Solar PV System Performance Assessment Guideline*. San Jose, CA: San Jose State University.

International Electrotechnical Commission (IEC). 2009. *Grid Connected Photovoltaic Systems—Minimum requirements for system documentation, commissioning tests and inspection*. May. IEC 62446.

International Energy Agency (IEA). 2010. *PV Technology Roadmap*. http://www.iea.org/publications/freepublications/publication/name,3902,en.html (accessed 9 February 2014).

International Finance Corporation (IFC). 2012. *Utility Scale Solar Power Plants: A Guide for Developers and Investors*. February. http://www.ifc.org/wps/wcm/connect/topics_ext_content/ifc_external_corporate_site/ifc+sustainability/publications/publications_handbook_solarpowerplants (accessed 9 December 2013).

Kollins, K., B. Speer, and K. Cory. 2010. *Solar PV Project Financing: Regulatory and Legislative Challenges for Third-Party PPA System Owners*. National Renewable Energy Laboratory. February. http://www.nrel.gov/docs/fy10osti/46723.pdf (accessed 5 February 2014).

Lee, D. 2011. *A Direct Comparison between a Central Inverter and Micro inverters in a Photovoltaic Array*. Master of Science Thesis. Appalachian State University. December. http://libres.uncg.edu/ir/asu/f/Lee,%20David_2011_Thesis.pdf (accessed 31 January 2014).

Mayfield, R. 2008. String Theory: PV Array Voltage Calculations. *Home Power.* June/July. http://www.homepower .com/articles/solar-electricity/design-installation/string-theory (accessed 10 February 2014).

Morales, A. and A. Vitelli. 2013. Europe's Carbon Emissions Market Is Crashing. 28 March. http://www .businessweek.com/articles/2013-03-28/europes-carbon-emissions-market-is-crashing (accessed 25 April 2014).

National Renewable Energy Laboratory (NREL). 2013. Best Research-Cell Efficiencies. http://www.nrel.gov/ncpv/ images/efficiency_chart.jpg (accessed 25 April 2014).

National VET. 2012. Derating Modules. Harness the Sun. https://nationalvetcontent.edu.au/alfresco/d/d/ workspace/SpacesStore/4e1e0e97-bc55-4c44-a692-4d31f4d0d934/13_02/content_sections/learn_about/08 _solar_page_002.htm (accessed 17 January 2014).

North American Board of Certified Energy Practitioners (NABCEP). 2005. *Study Guide for Photovoltaic System Installers and Sample Examination Questions.* August. http://www.brooksolar.com/files/NABCEP_Study _Guide-Revised_Version_3_-_08_05-FINAL.pdf (accessed 6 February 2014).

Pless, S., M. Deru, P. Torcellini, and S. Hayter. 2005. *Procedure for Measuring and Reporting the Performance of Photovoltaic Systems in Buildings.* National Renewable Energy Laboratory Technical Report NREL/TP-550 -38603. October. http://www.nrel.gov/docs/fy06osti/38603.pdf (accessed 14 February 2014).

Quantum Solar Power. 2012. A Comparison of PV Technologies. http://quantumsp.com/en/solar-energy/ a-comparison-of-pv-technologies/ (accessed 25 April 2014).

Renewable Energy Policy Network for the 21st Century (REN21). 2013. *Renewable Global Status Report 2013.* http:// www.ren21.net/REN21Activities/GlobalStatusReport.aspx (accessed 10 February 2014).

Sethuraman, D. and N. Pearson. 2011. Carbon Credits Becoming "Junk" before 2013 Ban Closes Door: Energy Markets. 8 December. http://www.bloomberg.com/news/2011-12-06/carbon-credits-becoming-junk-before -2013-ban-closes-door-energy-markets.html (accessed 25 April 2014).

Stoffel, T., D. Renné, D. Myers, S. Wilcox, M. Sengupta, R. George, and C. Turchi. 2010. *Concentrating Solar Power: Best Practices Handbook for the Collection and Use of Solar Resource Data.* National Renewable Energy Laboratory. Technical Report NREL/TP-550-46475. September. http://www.nrel.gov/docs/fy10osti/47465.pdf (accessed 20 December 2014).

United Nations Framework Convention on Climate Change (UNFCCC). 2012. The Doha Climate Gateway.http:// unfccc.int/key_steps/doha_climate_gateway/items/7389.php (accessed 24 April 2014).

United States Department of Energy (US DOE). 2006. Balance of System. http://web.archive.org/web/ 20080504001534/http:/www1.eere.energy.gov/solar/bos.html (accessed 5 January 2014).

US DOE. 2011. *Solar Powering Your Community: A Guide for Local Governments.* Solar Market Transformation Program. 2nd Edition. January. http://opr.ca.gov/s_renewableenergy.php (accessed 10 December 2013).

Virtuani, A., D. Pavanello, and G. Friesen. 2010. *Overview of Temperature Coefficients of Different Thin Film Photovoltaic Technologies.* In 25th European Photovoltaic Solar Energy Conference and Exhibition/5th World Conference on Photovoltaic Energy Conversion, pp. 6–10. Conference paper. http://www2.isaac.supsi.ch/ISAAC/Pubblicazioni/ Fotovoltaico/Conferences/Valencia%20(Spain)%20-%2025%20EU%20PVSEC%20-%20September%20 2010/4av3.83%20overview%20of%20temperature%20coefficients%20of%20different%20thin%20film%20 photovoltaic%20technologies%20(a.%20virtuani).pdf

World Energy Council. 2007. 2007 Survey of Energy Resources. September. http://www.worldenergy.org/ publications/survey_of_energy_resources_2007/default.asp (accessed 20 December 2013).

ANNEX 1

ADB Rooftop Solar Project Process

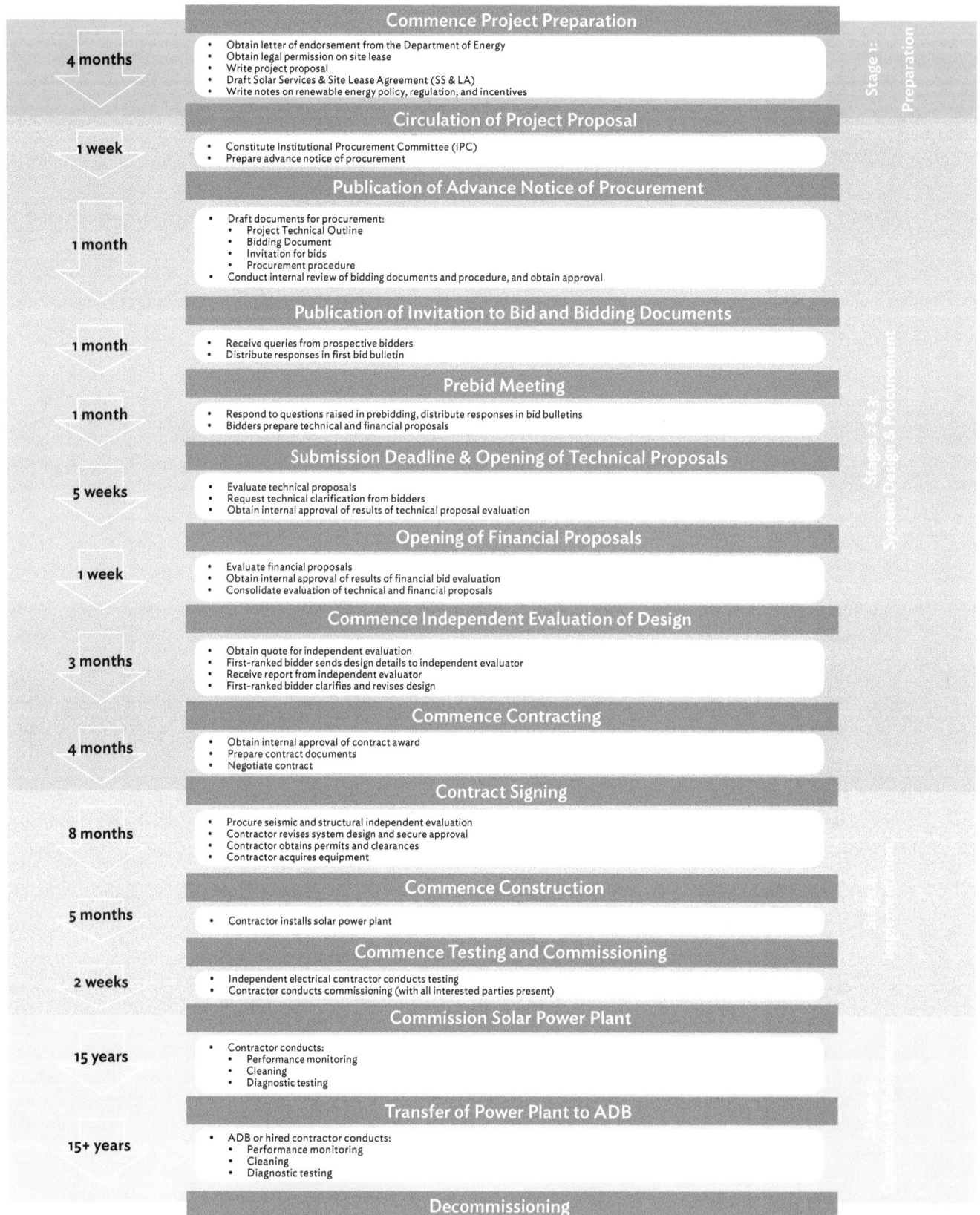

4 months

Commence Project Preparation

- Obtain letter of endorsement from the Department of Energy
- Obtain legal permission on site lease
- Write project proposal
- Draft Solar Services & Site Lease Agreement (SS & LA)
- Write notes on renewable energy policy, regulation, and incentives

1 week

Circulation of Project Proposal

- Constitute Institutional Procurement Committee (IPC)
- Prepare advance notice of procurement

1 month

Publication of Advance Notice of Procurement

- Draft documents for procurement:
  - Project Technical Outline
  - Bidding Document
  - Invitation for bids
  - Procurement procedure
- Conduct internal review of bidding documents and procedure, and obtain approval

1 month

Publication of Invitation to Bid and Bidding Documents

- Receive queries from prospective bidders
- Distribute responses in first bid bulletin

1 month

Prebid Meeting

- Respond to questions raised in prebidding, distribute responses in bid bulletins
- Bidders prepare technical and financial proposals

5 weeks

Submission Deadline & Opening of Technical Proposals

- Evaluate technical proposals
- Request technical clarification from bidders
- Obtain internal approval of results of technical proposal evaluation

1 week

Opening of Financial Proposals

- Evaluate financial proposals
- Obtain internal approval of results of financial bid evaluation
- Consolidate evaluation of technical and financial proposals

3 months

Commence Independent Evaluation of Design

- Obtain quote for independent evaluation
- First-ranked bidder sends design details to independent evaluator
- Receive report from independent evaluator
- First-ranked bidder clarifies and revises design

4 months

Commence Contracting

- Obtain internal approval of contract award
- Prepare contract documents
- Negotiate contract

8 months

Contract Signing

- Procure seismic and structural independent evaluation
- Contractor revises system design and secure approval
- Contractor obtains permits and clearances
- Contractor acquires equipment

5 months

Commence Construction

- Contractor installs solar power plant

2 weeks

Commence Testing and Commissioning

- Independent electrical contractor conducts testing
- Contractor conducts commissioning (with all interested parties present)

15 years

Commission Solar Power Plant

- Contractor conducts:
  - Performance monitoring
  - Cleaning
  - Diagnostic testing

15+ years

Transfer of Power Plant to ADB

- ADB or hired contractor conducts:
  - Performance monitoring
  - Cleaning
  - Diagnostic testing

Decommissioning

Stage 1: Preparation

Stages 2 & 3: System Design & Procurement

58 Source: ADB.

ANNEX 2

ADB Rooftop Solar Project Permits and Clearances

Permit/Clearance	Barangay Clearance for Renovation
Issuer	Barangay Wack-Wack Greenhills East
Details	Clearance from the *barangay*ᵃ is required to obtain the Locational Clearance from the City of Mandaluyong. The project was classified as a building renovation as it was installed on an existing facility.

Application Requirements	• Application form, signed by the company's authorized representative
	• Architectural plans (7 sets):
	  – Location/Vicinity plan
	  – Site development plan
	  – Floor plan
	  – Plan/Details
	    » Front, two sides, rear elevation
	    » Longitudinal or cross-section
	    » Roof framing
	    » Perimeter fence/wall
	  – Engineering design drawings
	    » Structural plans
	    » Electrical plans
	    » Topographic survey plan
	• Certified true copy of Transfer Certificate of Title (TCT) for construction location
	• Certification of rooftop leasing arrangements
	• Board resolution authorizing a signatory to transact construction
	• List of contact persons authorized to receive communications
	• Activity Clearance from Ortigas Center Association
	• Certificate of Non-Coverage

Processing Time	18 working days
Contact Details	Barangay Wack-Wack Greenhills East, 1507 & 1509 Princeton Street corner Shaw Boulevard, Mandaluyong City
Permit/Clearance	**Certificate of Non-Coverage (CNC)**
Issuer	Department of Environment and Natural Resources (DENR)
Details	A solar project is generally considered as "green" and environment-friendly, and unlikely to cause adverse environmental impacts. This certificate grants exemption from the Philippine Environmental Impact Assessment system.

Application Requirements	• Application form, signed by the company's authorized representative
	• Attachments:
	  – Project description
	  – Building location and area covered
	  – Building layout and map
	  – Site development plan with timelines
	  – Site photo of each side of the building
	  – Mayor's Permit
	  – Securities and Exchange Commission (SEC) Certificate
	  – Certification of rooftop leasing arrangements
	• Barangay Clearance for Renovation

continued on next page

table continued

Permit/Clearance	Certificate of Non-Coverage (CNC)
Processing Time	14 working days
Contact Details	DENR, Environmental Management Bureau, DENR Compound, Visayas Avenue, Diliman, Quezon City. +632 927-1517, +632 928-3782, www.emb.gov.ph

Permit/Clearance	Building Permit for Renovation
Issuer	City of Mandaluyong
Details	Included as part of the Mayor's Permit

Application Requirements	• Application form, signed by:
	  – Company's authorized representative
	  – Architect/Civil engineer for plans and specifications
	  – Architect/Civil engineer in-charge of construction
	  – Building owner
	• Architectural plans (5 sets):
	  – Site development and location
	  – Architectural plans and specifications
	  – Structural designs and specifications
	  – Electrical plans and specifications
	<u>Post-permit issuance:</u>
	• Certificate of Completion signed by:
	  – Designing architect/Civil engineer
	  – Architect/Civil engineer in-charge of construction, no later than 7 days from completion of project
	• Certificate of Occupancy prior to commencement of operations

Processing Time	21 working days
Contact Details	N/A

Permit/Clearance	Electrical Permit
Issuer	City of Mandaluyong
Details	Included as part of the Mayor's Permit

Application Requirements	• Application form signed by:
	  – Company's authorized representative
	  – Professional electrical engineer for plans and specifications
	  – Professional electrical engineer in-charge of construction
	• Electrical plans and specifications (5 sets)
	<u>Post-permit issuance:</u>
	• Certificate of Completion signed by the electrical practitioner in-charge of installation, no later than 7 days from completion of the installation, with as-built plan
	• Certificate of Final Electrical Inspection prior to on-grid operation

Processing Time	6 working days
Contact Details	City of Mandaluyong, Office of the Building Official, Office of the City Engineer, Electrical Section +632 532-5001

Permit/Clearance	Locational Clearance
Issuer	City of Mandaluyong
Details	Zoning clearance for the building permit that ensures the construction conforms to the city's comprehensive land use plan and zoning ordinances

continued on next page

table continued

Permit/Clearance	Locational Clearance
Application Requirements	• Notarized locational clearance application form • Barangay Clearance • Transfer certificate of title • Latest real property tax receipt (certified true copy of the tax declaration of real property can be obtained from the City Assessor's Office) • Architectural plan
Processing Time	5 working days
Contact Details	Zoning Division, City Planning and Development Office (CPDO), City of Mandaluyong
Permit/Clearance	**Fire Safety Evaluation Clearance**
Issuer	Mandaluyong City Fire Station
Details	Confirms that the renovations meet the Fire Code of the Philippines during the construction period. It is processed simultaneously with the building permit application.
Application Requirements	• Plans and specifications • Fire and Life Safety Assessment Report (FALAR) 1 (3 sets) (This documents the fire and life safety features of the building, and must be prepared by an architect who is certified by the Bureau of Fire Protection to do so.)
Processing Time	15 working days
Contact Details	Mandaluyong City Fire Station, National Capital Region Fire District IV, Bureau of Fire Protection, Department of the Interior and Local Government
Permit/Clearance	**Aviation Authority Height Clearance Permit**
Issuer	Civil Aviation Authority of the Philippines (CAAP)
Details	The solar farm at ADB's rooftop is to be located below the building's helipad, which means that there was no need to secure a Height Clearance Permit or any clearance from CAAP.
Application Requirements	ADB sent an enquiry letter to CAAP's Flight Standards Inspectorate Service to confirm that no permit or clearance was required.
Processing Time	8 working days
Contact Details	CAAP, Department of Transportation and Communications, MIA Road cor. Ninoy Aquino Avenue, Pasay City, Metro Manila. +632 879-9286, www.caap.gov.ph
Permit/Clearance	**Activity Clearance**
Issuer	Ortigas Center Association
Details	Advises the Ortigas Center Association of the planned construction, valid for 2 weeks, for a minimal fee of P10
Application Requirements	N/A
Processing Time	6 working days
Contact Details	Ortigas Center Association, Inc., OCAI Building, Garnet Road, Ortigas Center, Pasig City

[a] The *barangay* is the basic political unit in the Philippines.

Source: ADB.

ANNEX 3

Policy, Regulation, and Incentives for Solar Rooftops in the Philippines

A3.1 Renewable Energy Act 2008 (Republic Act No. 9513)

Signed into law in December 2008, Republic Act No. 9513 or the Renewable Energy Act of 2008 (the Act) promotes the development, utilization, and commercialization of renewable energy resources in the country. The Act calls for the state to:

- accelerate the deployment of renewable energy technologies for the purpose of enhancing energy security;
- institutionalize renewable energy development;
- promote the use of renewable energy in commercial applications through both fiscal and nonfiscal incentives;
- promote renewable energy as part of a strategy to achieve economic growth while protecting health and the environment; and
- establish the necessary infrastructure and mechanisms necessary to carry out these goals.

The Act additionally mandates the Department of Energy (DOE) to develop the Implementing Rules and Regulations of the Act. Rules and regulations pertaining to rooftop solar development include rules for awarding certificates for feed-in tariff eligibility,[1] registering as a renewable energy developer;[2] adopting templates for renewable energy service contracts,[3] and establishing timelines and processes for reviewing renewable energy service contracts.[4] The DOE publishes these regulations on its website.

A3.2 Incentives

The DOE has divided regulations for incentivizing renewable energy development into two categories: (i) fiscal incentives and (ii) "other incentives and privileges." Fiscal incentives provide direct financial benefits in the form of tax reductions, exemptions, carry-overs, etc. Other incentives also provide financial benefits to the developer.

To provide an example of what some of these incentives look like, Tables A3.1 and A3.2 list fiscal and other incentives, as they existed as of February 2014.

---

[1] See, e.g., Resolution Approving the Feed-In Tariff Rates (Resolution No. 10, Series of 2012); see also Guidelines for the Selection Process of Renewable Energy Projects under Feed-in Tariff System and the Award of Certificate for Feed-in Tariff Eligibility (DOE Department Circular No. DC2013-05-0009).

[2] Guidelines Governing a Transparent and Competitive System of Awarding Renewable Energy Service/Operating Contracts and Providing for the Registration Process of Renewable Energy Developers (DOE Department Circular No. DC2009-05-0011).

[3] Adopting Policies in Relation to the Processing of Renewable Energy Service Contracts and Mandating the Adoption of the Revised Templates for Renewable Energy Service Contracts (DOE Department Order No. DO2013-08-0011).

[4] Adopting the Revised Evaluation Process Flow and Timelines of Renewable Energy Service Contracts (RESC) and Mandating the Adoption of the Milestone Approach (DOE Department Order No. DO2013-10-0018).

Table A3.1: Fiscal Incentives for Renewable Energy Projects and Activities

Fiscal Incentive	Description
Income Tax Holiday	Developers are eligible to receive a full exemption from income taxes for 7 years from the start of commercial operations. Developers may continue to avail of a tax holiday after 7 years upon making an additional investment to existing operations. However, the maximum amount of time any project is eligible for is 21 years. Developers must register and have the Department of Energy approve of the additional investments.
Corporate Tax Rate	Renewable energy developers pay a corporate tax of 10% on net taxable income under two circumstances: (1) after availing of the income tax holiday, provided that they pass on the savings to end-users in the form of lower power rates; and (2) if the renewable energy developer acquired, operated, or administered the existing renewable energy facilities that have been in operation for more than 7 years.
Government Share	The government is entitled to a share of 1% of the gross income of solar energy developers. However, projects for own use of power and microscale (≤ 100 kW installed capacity) projects for noncommercial use that have registered with the DOE are exempt from this payment.
Accelerated Depreciation	A renewable energy developer is eligible to receive accelerated depreciation if the developer fails to receive an income tax holiday before becoming fully operational; however, once a renewable energy developer receives accelerated depreciation, the developer is no longer eligible for an income tax holiday. It allows the developer to depreciate the plant, machinery, and equipment at twice the normal depreciation rate.
Duty-Free Importation of Renewable Energy Machinery, Equipment, and Materials	Within the first 10 years from the issuance of a Certificate of Registration to a renewable energy developer, the importation of machinery and equipment is exempt from importation duties. Exemption is conditional on the developer actually needing the renewable energy equipment and requires authorization.
Special Realty Tax Rates on Equipment and Machinery	Realty and other taxes on civil works, equipment, and other improvements by a registered renewable energy developer actually and exclusively used for renewable energy facilities shall not exceed 1.5% of their original cost less accumulated normal depreciation or net book value.
Net Operating Loss Carry-Over	Any losses from the first 3 years from the start of commercial operation shall be carried over as a deduction from gross income. The loss can be carried over for the next 7 consecutive taxable years immediately following the year of such loss.
Zero Percent Value-Added Tax (VAT) Rate	The sale of fuel generated from renewable energy, purchase of local goods and services needed by renewable energy development are subject to a 0% value-added tax.
Tax Exemption of Carbon Credits	All proceeds from the sale of carbon emission credits shall be exempt from any and all taxes.
Tax Credit on Domestic and Capital Equipment and Services	Renewable energy developers purchasing domestic machinery, equipment, and materials receive a tax credit equivalent to the value-added tax and customs duties that would have been paid if they had been imported.

kW = kilowatt.

Source: Department of Energy (DOE) Department Circular No. DC2009-05-0008, Part III, Rule 5, Section 13; The Renewable Energy Act of 2008, Chapter V, Section 13.

Table A3.2: Other Incentives and Privileges

Other Incentives and Privileges	Description
Tax Rebate for Purchase of Renewable Energy Components	The Department of Finance is to establish tax rebates for the purchase of renewable energy equipment for residential, industrial, or community use.
Exemption from the Universal Charge	A universal charge includes the charge imposed for stranded costs. All consumers shall be exempted from paying a universal charge if the power or electricity generated through the renewable energy system is consumed by the generators themselves; and/or distributed free of charge in an off-grid area.
Payment of Transmission Charges	A registered renewable energy developer producing power and electricity from an intermittent renewable energy resource may opt to pay the transmission and wheeling charges at a cost equivalent to the average per kilowatt-hour rate of all other electricity transmitted through the grid.
Priority and Must Dispatch for Intermittent Renewable Energy Resource	Qualified and registered renewable energy generating units shall be considered "must dispatch" based on available energy and shall enjoy the benefit of priority dispatch.
Preferential Financing for Solar Energy	The Renewable Energy Act 2008 mandates Philippine Government Financial Institutions to provide preferential financial packages for the development, utilization, and commercialization of renewable energy projects that are duly recommended and endorsed by the Department of Energy.

Source: Department of Energy (DOE) Department Circular No. DC2009-05-0008; Part III, Rule 5, Section 17.

A3.2.1 Feed-in Tariff

Effective as of July 2012, the Energy Regulatory Commission (ERC) approved a guaranteed, fixed feed-in tariff (FIT) of P9.68 ($0.225 per kilowatt-hour) for solar projects.[5] The FIT has a duration of 20 years and a degression rate of 6% after the first year. The ERC may review and readjust the FIT every 3 years, or when installation targets set by the DOE are met.

A3.2.2 Net Metering

In May 2013, the ERC established rules for net metering in the Philippines in consultation with the National Renewable Energy Board (NREB) and electric power industry participants (ERC Resolution No. 09, Series of 2013). Customers are eligible for the net metering program provided they are in good standing with regards to paying electricity bills and own systems with only up to 100 kW in capacity. Customers must also comply with relevant national codes and standards, including the Net-Metering Interconnection Standards.[6]

Any excess electricity generated by the rooftop solar power system and exported to the distribution utility earns credits (in Philippine pesos), which customers may credit against their electric bill. The distribution utility is entitled to any Renewable Energy Certificates from this electricity and can use it in compliance with Renewable Portfolio Standards. The Renewable Energy Market, which was established by the DOE and regulated by the Philippine Energy Market Corporation, facilitates the trade of Renewable Energy Certificates between power producers and the distribution utility.[7]

---

[5] ERC Resolution No. 10, Series of 2012.
[6] ERC Resolution No. 09, Series of 2013.
[7] DOE, Department Circular No. DC2009-05-0008, Rule 3: Renewable Energy Market, Sections 10-11.

A3.3   Eligibility

To be entitled to incentives and privileges under the Renewable Energy Act of 2008, developers of projects must meet the DOE's and the Board of Investments' registration and accreditation requirements.[8] The DOE delegated the Renewable Energy Management Bureau to administer the registration and accreditation of developers and projects.

Developers must be either a Filipino national or an incorporated Filipino company, with at least 60% Filipino ownership in the company's capital. The company must also be registered with the Securities and Exchange Commission (SEC).

As of March 2014, regulations treat all rooftop solar generators as an "energy generator," with the exception of projects for "own use," and noncommercial projects under 100 kilowatts-peak (kWp) (as will be discussed). DOE regulations require solar rooftop generators, as energy generators, to enter into a Solar Energy Service Contract with the Government of the Philippines and deliver electricity to the grid. Eligibility for incentives and privileges is built into the contracting process, which is described in Section A3.4.

These regulations also treat third-party generators as energy generators, even if the third party intends to sell electricity to only the building owner. In practice, third-party generators need not register or enter into a Solar Energy Service Contract for their power purchase agreements or solar lease agreements to be valid, but they forfeit fiscal incentives and privileges.

For projects where renewable electricity generation is for own use and for noncommercial projects under 100 kWp, the DOE has a simplified registration process.[9]

Projects for own use are those in which the owner of the building or facility purchases the solar technology, and all of the electricity generated is used within the building or facility and not connected to the grid. In this case, there is no limit to system capacity.

Microscale projects for noncommercial use is similar except they have a capacity restriction of 100 kWp. This category is intended for donor-driven systems in small communities or public facilities such as schools, where the installers are paid a predefined price over time to get back their investment in the solar power system.

Projects with net metering are treated separately to the types of projects described above. The Deutsche Gesellschaft für Internationale Zusammenarbeit (GIZ), in conjunction with the NREB, has released a guide for net metering in the Philippines.[10] This guide provides details of calculations of charges and credits for net metering, and the processes that project developers must go through to set up net metering for their rooftop solar project.

A3.4   Solar Energy Service Contract Process

A Solar Energy Service Contract is a service agreement with the Government of the Philippines (through the President or the DOE) that gives the solar energy developer the exclusive right to explore, develop, or utilize a particular area.

---

[8]   DOE Department Circular No. DC2009-05-0008.
[9]   DOE Department Circular No. DC2009-07-0011.
[10]   National Renewable Energy Board and GIZ. 2013. *Net-Metering Reference Guide: How to Avail Solar Roof Tops and Other Renewables below 100 kW in the Philippines.* November 2013. Deutsche Gesellschaft für Internationale Zusammenarbeit (GIZ) GmbH. http://climatechange .denr.gov.ph/index.php?option=com_docman&task=doc_download&gid=125&Itemid=4 (accessed 19 February 2014).

This section details the process that project developers must go through to secure a Solar Energy Service Contract with the DOE, including application requirements, procedure for awarding of contracts, and registration procedure. Information provided in this section is based on the DOE's guidelines on renewable energy service and operating contracts.[11]

A3.4.1 Application Process for Commercial Projects

Contract Stages. The contracts cover two project stages:

(i) the Pre-Development Stage, which covers the preliminary assessment and feasibility study, and up to financial closing of the solar energy project; and
(ii) the Development/Commercial Stage, which covers the development, production, or utilization of renewable energy resources, including the construction and installation of relevant facilities, and up to the operation phase of the renewable energy facilities.

Blocking System. In the blocking system, the territory of the Philippines is subdivided into square blocks of 81 hectares each,[12] and each block pertains to a single contract area. Project developers should first verify that the block or blocks in which their project falls has not already been allocated to another renewable energy service contract. Check with the map on file with the Information Technology and Management Services (ITMS) unit of the DOE.

Once the boundaries of a project site are known, the project developers submit a map with the boundaries of their contract site. The remainder of the block is then available for other renewable energy service contracts.

Application for Solar Energy Service Contract. To apply for a Solar Energy Service Contract, project developers should submit an application to the Department of Energy. The application should include the following:[13]

A. **Legal Requirements**
1. Individual or single proprietorship
   a. Birth Certificate—duly authenticated by National Statistics Office (NSO)
   b. Business Permit—certified true copy
   c. Department of Trade and Industry (DTI) Registration (if applicable).

2. Corporation/Joint venture/Consortium
   a. Securities and Exchange Commission (SEC) Registration—SEC-certified
   b. By-laws and Articles of Incorporation—SEC-certified
   c. Certification authorizing its representative to negotiate and enter into Renewable Energy Contract with the DOE
   d. Business Permit
   e. Controlling stockholders and percentage of their holdings
   f. Organizational chart of the company
   g. Parent/Subsidiary/Affiliates (if applicable)
   h. Company profile

[11] *Guidelines Governing a Transparent and Competitive System of Awarding RE Service/Operating Contracts and Providing for the Registration Process of RE Developers (Department of Energy [DOE] Circular No. DC2009-07-0011), and Adopting Policies in Relation to the Processing of Renewable Energy Service Contracts and Mandating the Adoption of the Revised Templates for Renewable Energy Service Contracts* (DOE Department Order No. DO2013-08-0011).
[12] The meridional blocks are sized 0.5 minutes of latitude by 0.5 minutes of longitude using the Philippine Reference System (PRS) Geographic Projection and Datum of 1992.
[13] Updated from DOE. 2013. *Checklist of Requirements (Renewable Energy Service/Operating Contract under R.A. 9513)*. Brochure. http://www.doe.gov.ph/doe_files/pdf/Researchers_Downloaded_Files/Brochures/Checklist_of_Requirements_RE_SC.pdf (accessed 5 December 2013).

B. **Technical Requirements**
1. Track record or experience
2. Work program with financial commitment per activities
3. Curriculum vitae of management and technical personnel
4. List of technical consultants with corresponding contract between the developer and consultants showing their respective qualifications
5. List of existing company-owned and leased equipment appropriate for the renewable energy project with corresponding description

C. **Financial Requirements**
1. Audited Financial Statement for the last 2 years and unaudited Financial Statement if the filing date is 3 months beyond the date of the submitted Audited Financial Statement
2. Bank certification to substantiate the cash balance (exact amount in words and numbers)
3. Projected cash flow statement for 2 years
4. For a newly organized or subsidiary corporation with insufficient funds to finance the proposed work program, it shall submit an Audited Financial Statement and duly certified and/or notarized guarantee or Letter of Undertaking/Support from its parent company or partners to fund the proposed work program. In the case of a foreign parent-company, the Audited Financial Statement and the guarantee or Letter of Undertaking/Support shall be duly authenticated by the Philippine Consulate Office that has consular jurisdiction over said parent-company.

D. **Other Requirements**
1. Letter of Intent/Application
2. Duly accomplished Renewable Energy Contract Application Form (DOE Department Circular No. DC2009-07-0011, Annex A)
3. Map showing the area of application, conforming with the DOE Blocking System
4. Application/Processing fees[14]
5. Draft Solar Energy Service Contract (as in DOE Department Order No. DO2013-08-0011)

Figure A3.1 shows the registration and evaluation process and timelines. Except in extreme circumstances, the DOE will evaluate the application within 45 working days (DOE Department Order No. DO2013-10-0018).

If the application is successful, both parties will sign the contract and the project enters the predevelopment stage, which covers the preliminary assessment and feasibility study up to the financial closing of the project.

A3.4.2 Conversion to Development/Commercial Stage

If the project developer finds the project to be commercially feasible, it should apply for conversion from Pre-Development Stage to Development/Commercial Stage. The application should include the following:[15]

A. Letter of Declaration of Commerciality declaring the renewable energy project is commercially feasible and viable; and
B. Feasibility study and/or detailed engineering design of the renewable energy project with the following corresponding documents:
1. Resolution of support from host communities and host municipality/ies
2. Proof of Public Consultation

---

[14] Fees and charges for renewable energy service contract applications are posted online at http://www.doe.gov.ph/issuances/fees-and-charges
[15] Updated from DOE. 2013. *Checklist of Requirements (Renewable Energy Service/Operating Contract under R.A. 9513)*. Brochure. http://www.doe.gov.ph/doe_files/pdf/Researchers_Downloable_Files/Brochures/Checklist_of_Requirements_RE_SC.pdf (accessed 5 December 2013).

Figure A3.1: Evaluation Process Flow and Timelines of Renewable Energy Service Certificate Applications (Direct Negotiation)

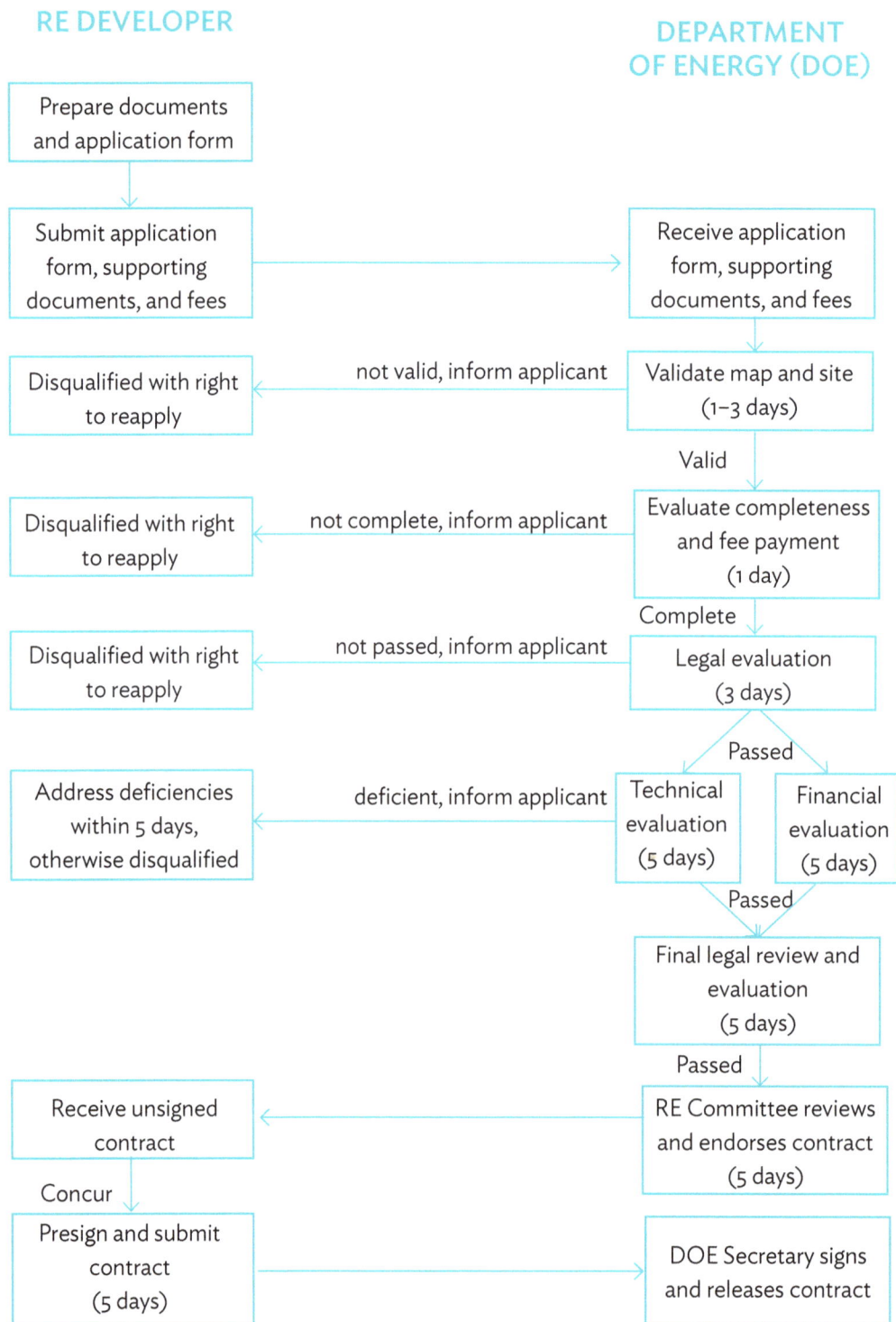

RE DEVELOPER

DEPARTMENT OF ENERGY (DOE)

Prepare documents and application form

Submit application form, supporting documents, and fees → Receive application form, supporting documents, and fees

Disqualified with right to reapply ← not valid, inform applicant ← Validate map and site (1–3 days)

Valid

Disqualified with right to reapply ← not complete, inform applicant ← Evaluate completeness and fee payment (1 day)

Complete

Disqualified with right to reapply ← not passed, inform applicant ← Legal evaluation (3 days)

Passed

Address deficiencies within 5 days, otherwise disqualified ← deficient, inform applicant ← Technical evaluation (5 days) | Financial evaluation (5 days)

Passed

Final legal review and evaluation (5 days)

Passed

Receive unsigned contract ← RE Committee reviews and endorses contract (5 days)

Concur

Presign and submit contract (5 days) → DOE Secretary signs and releases contract

RE = renewable energy.

Source: Modified from DOE Department Order No. DO2013-10-0018.

3. Any form of legal documents showing the consent of the landowner if the project falls on a private land
4. Department of Environment and Natural Resources (DENR) Permits:
    a. Environmental Impact Study
    b. Environmental Compliance Certificate (BCC) or Certificate of Non-Coverage (CNC)
    c. Forest Land Use Agreement (FLAg/Special Land Use Agreement (SLUP) for area applied in public domain
5. National Commission on Indigenous Peoples (NCIP): Free and Prior Informed Consent (FPIC)/ Certification of Pre-Condition or Certificate of Non-Overlap
6. National Transmission Corporation (TRANSCO):
    a. Grid System Impact Study
    b. Interconnection Agreement, if applicable
7. Energy (Electricity) Sales Agreement
8. Other clearances from other concerned agencies (i.e., Maritime Industry Authority [MARINA], Bureau of Fisheries and Aquatic Resources [BFAR], Philippine Navy, Philippine Coast Guard, etc.)
9. Proof of Financial Closing
10. Final area for development (geographical coordinates/PRS92)
11. Payment of corresponding application/processing fee
12. Draft Development/Commercial Renewable Energy Contract

The DOE will then evaluate the application and issue a Certificate of Confirmation of Commerciality to those which qualify. The project will then be in the Development/Commercial Stage, and construction of the solar energy system can commence.

A3.4.3   Feed-in Tariff

Projects eligible for the first round of the FIT are the first 50 megawatts (MW) of solar projects that are issued a Certificate of Endorsement for FIT Eligibility from the time the approved rates were set (July 2012). Eligibility for the FIT does not affect eligibility for Renewable Energy Certificates, which are used in conjunction with the Renewable Portfolio Standard mechanism (as will be discussed).

Payment is based on electricity delivered to the transmission and/or distribution network and excludes generation for own use.[16]

To receive a FIT payment, developers of solar power projects must undergo the following steps:

1. Hold a valid Solar Energy Service Contract.
2. Submit an application for conversion of the Solar Energy Service Contract from Pre-Development Stage to Development Stage, which
    a. includes a Declaration of Commerciality based on the approved FIT rate and a Work Plan; and
    b. indicates that the Declaration of Commerciality is based on the approved FIT rate.[17]
3. The DOE processes the Declaration of Commerciality and assesses compliance for conversion from Pre-Development Stage to Development Stage under the FIT System within 30 working days.
4. The DOE issues compliant projects with a Certificate of Confirmation of Commerciality, which serves as a notice to proceed to the construction phase. The DOE may issue an endorsement to the National Grid

---

[16]   ERC Resolution No. 15, Series of 2012.
[17]   If the project already has a Solar Energy Service Contract under the Development Stage, or if it is covered under a Renewable Energy Operating Contract, the developer may still apply for FIT eligibility as long as it is not bound to any contract to supply its generated energy to a distribution utility or consumer, and provides notarized proof or a declaration as such (DOE Department Circular No. DC2013-05-0009).

Corporation of the Philippines to conduct interconnection requirements (e.g., Grid Impact Study and Interconnection Agreement).

5. Upon Electromechanical Completion (i.e., 80% of the construction and development completed as defined in the engineering, procurement, and construction contract or work plan), the developer informs the DOE, who will inspect and validate the site (including the interconnection facility) within 15 working days.

6. Within 5 working days, the DOE nominates completed projects (with interconnection facilities fully in place) to the ERC as eligible for the Certificate of Compliance under the FIT System.

7. The developer informs the DOE upon successful plant commissioning, which the DOE validates. Within 15 working days from the validation, the DOE will issue a Certificate of Endorsement for FIT Eligibility for the plant to ERC.

The DOE maintains a monitoring board on its website to show the status of projects under each type of emerging renewable energy plants that apply for Conversion from Pre-Development to Development/Commercial Stage[18] and the Certificate of Confirmation of Commerciality.[19] Once the cumulative capacity of issued Certificates of Confirmation of Commerciality exceeds the installation target, the DOE and the NREB will review the targets and the FIT Scheme.

Once issued Certificates of Endorsement for FIT Eligibility reach the installation target, developers can either enter into bilateral agreements with a distribution utility or other off-taker, or export the generated power directly to the wholesale electricity spot market on a "must-dispatch" basis. They will be eligible for the Certificate of Endorsement for FIT Eligibility under the next FIT System regime.

The Philippine FIT applies to projects that achieve certain capacity factor levels, thus providing project developers with an additional incentive to design the most efficient projects.

A3.5   Projects for Own Use of Power

Projects that generate power only for use in the facility itself do not enter into a solar energy services contract with the government. They need only register with the DOE if they wish to avail of incentives. The DOE issues a Certificate of Registration to these projects once all of the following requirements are submitted and validated (DOE Department Circular No. DC2009-07-0011, Sec. 26):

(i)   letter of intent;
(ii) project description; and
(iii) proof of ownership of the facilities.

The blocking system does not apply to projects for own use of power.

---

[18]   See http://www.doe.gov.ph/feed-in-tarriff-monitoring-board/for-conversion-from-pre-development-to-development-commercial-stage
[19]   See http://www.doe.gov.ph/feed-in-tarriff-monitoring-board/with-certificate-of-confirmation-of-commerciality

A3.6   Microscale Projects for Noncommercial Use

The application process is simplified for microscale projects for noncommercial use. The developer should submit the following to the Renewable Energy Management Bureau:[20]

(i)   letter of intent;
(ii)   project description;
(iii)   work plan;
(iv)   local government endorsement/certification;
(v)   legal requirements as required for commercial contracts; and
(vi)   other proof of sustained operations.

---

[20]   DOE Department Circular No. DC2009-07-0011.

ANNEX 4
ADB Rooftop Solar Project Shading Analysis

This annex shows how the Asian Development Bank (ADB) conducted its shading analysis, given that ADB headquarters is particularly subject to shading from the southwest of the building. Showing ADB's shading analysis is meant to assist developers with managing possible shading surrounding their own solar projects, since it can seriously compromise solar system performance.

ADB's shading analysis involved obtaining and comparing results using two different methods.

The first method is called the spherical picture method. As will be described in more detail, it essentially involves taking photographs of the view of the sky that the solar array will experience. This gave a quick impression of the solar window of the sky without external data gathering, and included the effects of shading from all nearby structures and from all obstructions on the roof.

The second method utilizes simulation software, which models shading using a three-dimensional (3D) modeling tool that tracks the sun's path for the specific location of the site. After tracking, the software calculates the shading effect of the surrounding buildings based on the height, area, and orientation of each with respect to the location of the solar array.

In comparing results from both methods, the first method yielded a performance ratio of 1.7% higher than the second method, which is equivalent to a 1.23% higher annual energy production.

A4.1   Spherical Picture Method

The first method requires a digital camera with a fish-eye lens, a compass, a GPS device, and image processing software. ADB took pictures with the camera pointing directly upward from each of the locations of the roof shown in Figure A4.1, and processed the images to remove all but the obstructions from the images (Figure A4.2).

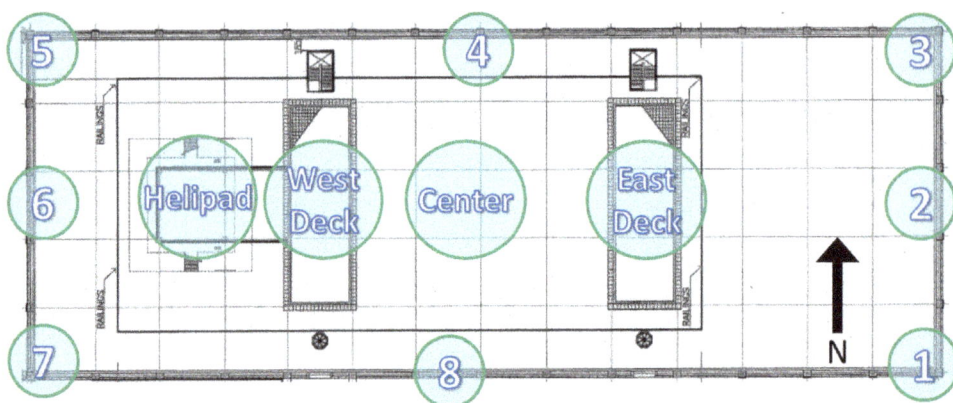

Figure A4.1:   Locations of Photographs for Shading Analysis

Source: ADB.

Figure A4.2: Processed Spherical Photo from Point 1

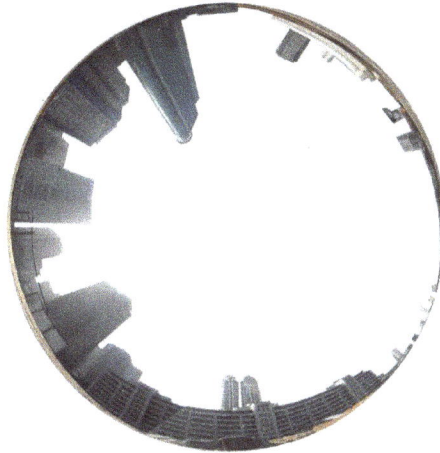

Source: ADB.

Figure A4.3: Solar Path Chart Developed for the ADB Rooftop Solar Power Project

Solar paths at ADB Sector C, (Lat. 14.6°N, long. 121.1°E, alt. 86 m)

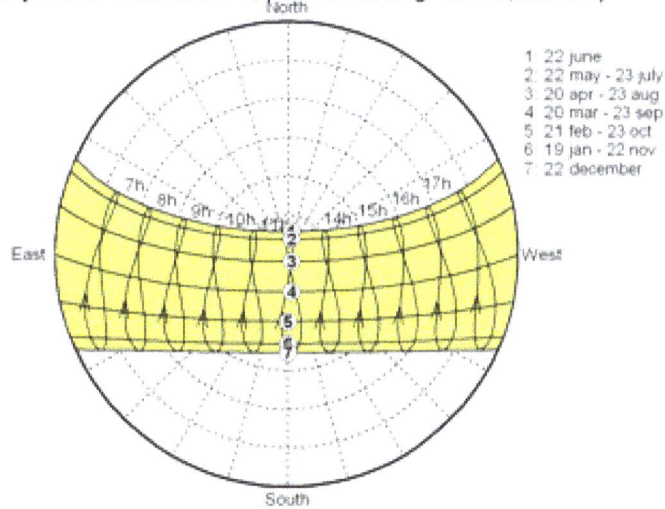

1 22 june
2 22 may - 23 july
3 20 apr - 23 aug
4 20 mar - 23 sep
5 21 feb - 23 oct
6 19 jan - 22 nov
7 22 december

Note: ADB's Headquarters is located on the northern hemisphere. In a solar path chart of this type, the sun's position is taken from the perspective of an observer facing the equator. Thus, east appears on the left and west appears on the right.

Source: ADB.

Next, ADB generated a solar path chart specifically for the latitude and longitude of the roof (Figure A4.3) using photovoltaic system design software.[1] This chart shows the path of the sun for the location at various times of the year.

The processed spherical photo was then overlaid on the solar path chart to determine shading (Figure A4.4).

----

[1] ADB used the software PVsyst, available at http://www.pvsyst.com

Figure A4.4: Spherical Picture Overlaid on the Sun Path, Used for Shading Charts

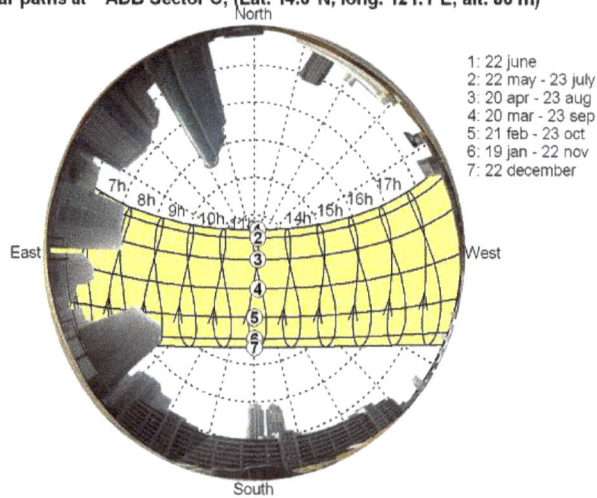

Solar paths at ADB Sector C, (Lat. 14.6°N, long. 121.1°E, alt. 86 m)

1: 22 june
2: 22 may - 23 july
3: 20 apr - 23 aug
4: 20 mar - 23 sep
5: 21 feb - 23 oct
6: 19 jan - 22 nov
7: 22 december

Note: ADB's Headquarters is located on the northern hemisphere. In a solar path chart of this type, the sun's position is taken from the perspective of an observer facing the equator. Thus, east appears on the left and west appears on the right.

Source: ADB.

Figure A4.5: Spherical Pictures Overlaid on the Sun Path Charts at Each Roof Location (Reversed East–West Direction)

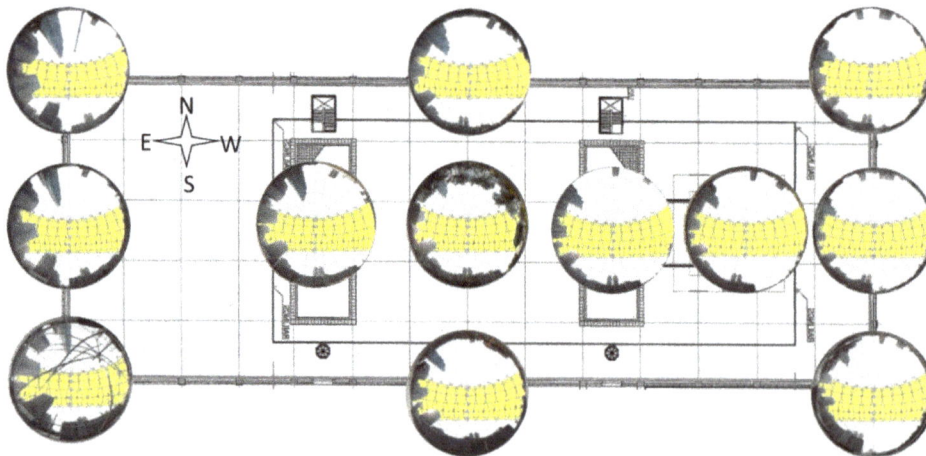

Note: ADB's Headquarters is located on the northern hemisphere. In a solar path chart of this type, the sun's position is taken from the perspective of an observer facing the equator. Thus, east appears on the left and west appears on the right.

Source: ADB.

Figure A4.5 shows the overlaid charts at each of the locations investigated.

ADB then manually translated the azimuth and angular height of the buildings onto the far horizontal shading diagram from the photovoltaic system design software, as shown in Figure A4.6.

Figure A4.6: Far Horizontal Shading Diagram for Point 1

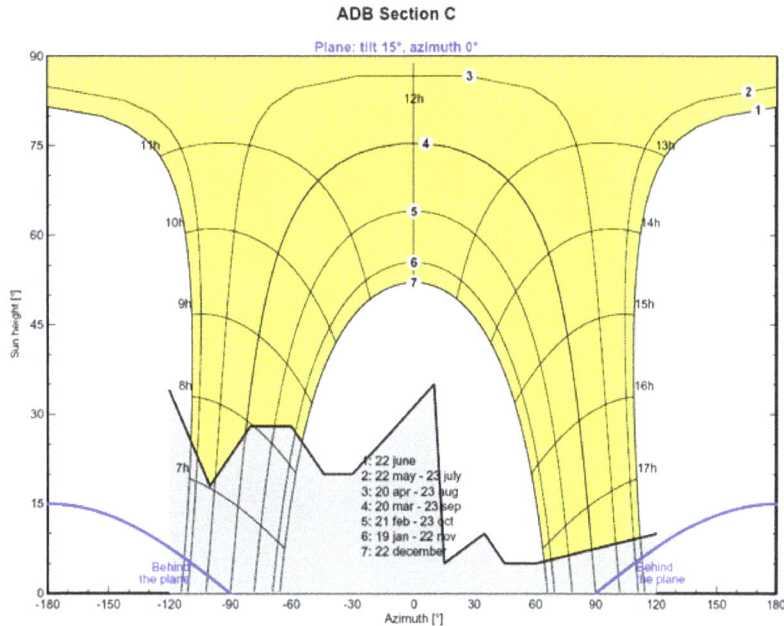

Source: ADB.

A4.2   Simulation Software Sketch-Up Method

The second method requires the physical dimensions of the roof and surrounding structures. These can be collected using city maps containing building information, from a physical survey, or using satellite images and the known number of floors in surrounding buildings.

ADB collected data on surrounding buildings (Table A4.1) and used a satellite image to determine the exact location, orientation, and dimensions of the buildings.

Table A4.1:   Buildings Surrounding ADB Headquarters

Name of Building	Number of Floors
Malayan Plaza	40
Oakwood Joy-Nostalg Center (new)	41
Jollibee Plaza	34
East Oak Galleria (new)	46
City Land Mega Plaza	40
Robinson Equitable Tower	45

Source: ADB.

ADB used the 3D modeling tool within the photovoltaic system design software to translate the collected building data, and also the size, tilt, location, and orientation of the proposed solar arrays into a 3D map, as shown in Figure A4.7. Figure A4.8 shows the shading diagram that the software automatically produced.

Figure A4.7: 3D Map of ADB Headquarters and Surroundings

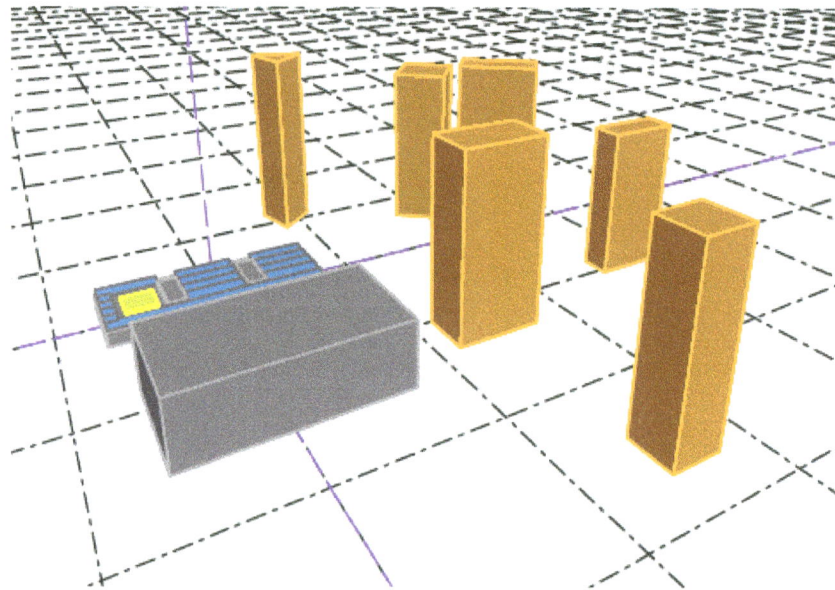

Source: ADB.

Figure A4.8: Iso-Shading Diagram Using 3D Map as Input

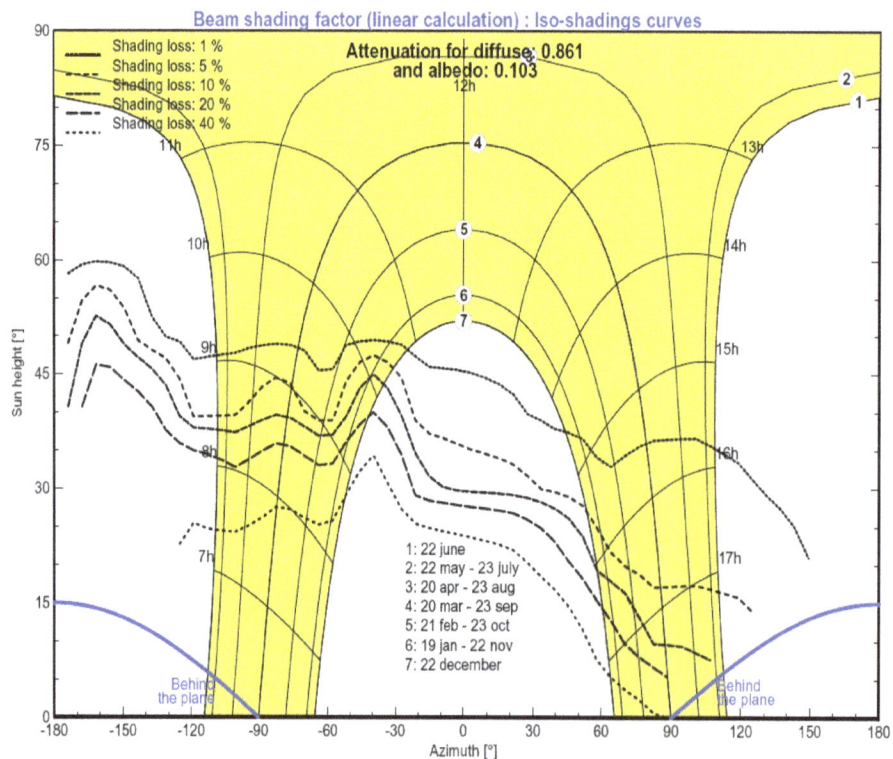

Source: ADB.

A4.3   Future Structures

ADB will construct two additional buildings in the future that can affect the shading of the solar array on the roof of Segment C (Facilities Block): a seven-story parking area and a third atrium. ADB added these structures to the shading model (Figure A4.9) to produce a shading map that takes them into account.

The new structures will cast shade from the southwest, as shown in Figure A4.10.

Figure A4.9:   3D Map of ADB Headquarters and Surroundings, Including Future Structures

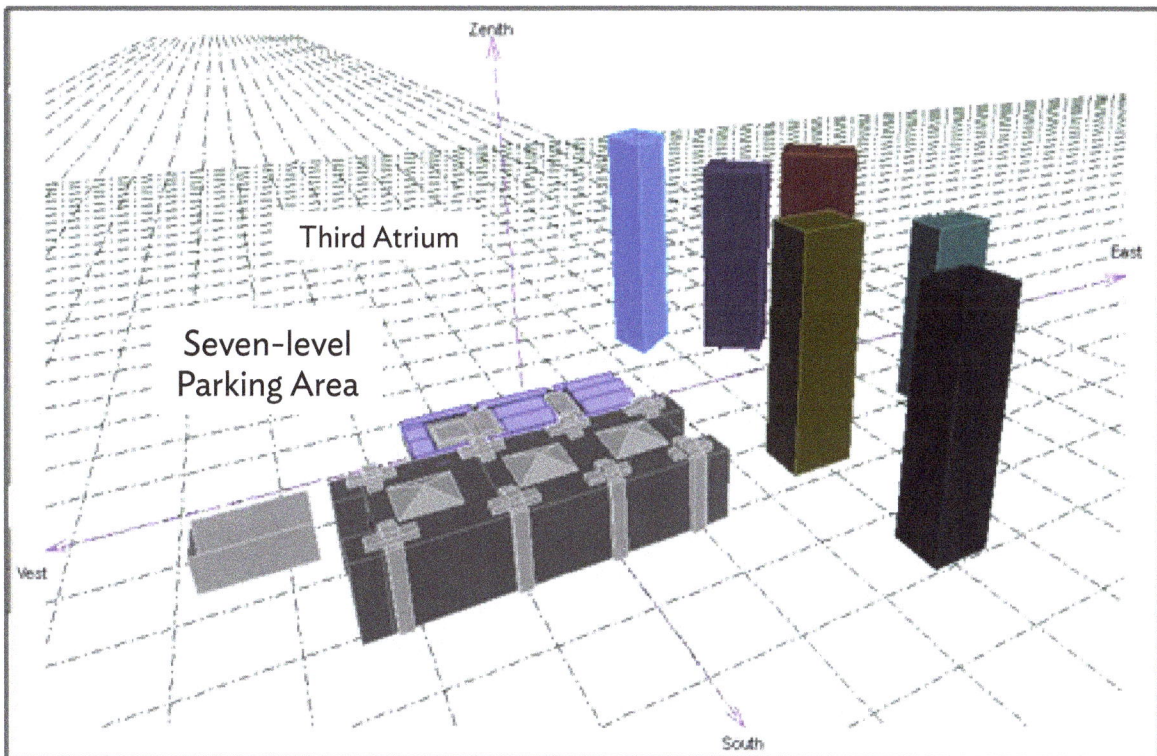

Source: ADB.

Figure A4.10: Increased Shading from Future Structures (in Orange Circle) Shown Using Iso-Shading Curves

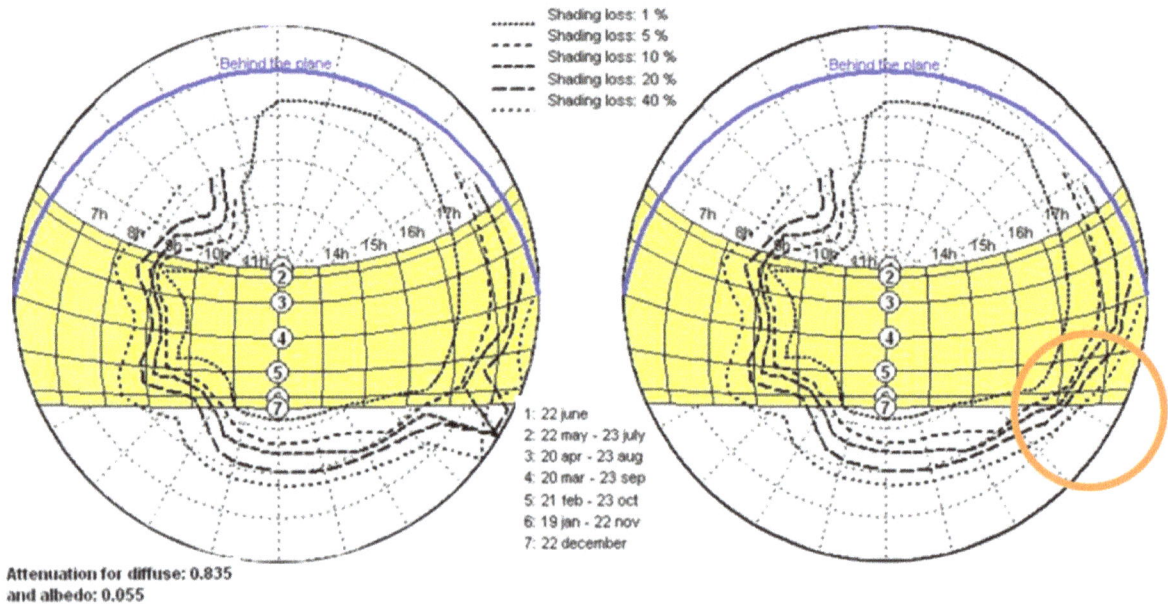

Note: ADB's Headquarters is located on the northern hemisphere. In a solar path chart of this type, the sun's position is taken from the perspective of an observer facing the equator. Thus, east appears on the left and west appears on the right.

Source: ADB.

ANNEX 5
ADB Rooftop Solar Project Technical Outline
Introductory Guidelines for the Project Developer[1]

These guidelines are intended to provide basic information to bidders regarding design processes, installation, trial test, and operation and maintenance (O&M) of the solar project. Further detailed information regarding the building can be requested directly from the Asian Development Bank (ADB) Facilities Manager individually.

The bidder shall present its own technical design and which will cover (but is not limited to) each one of the steps suggested within this document, in order to justify the claimed yearly energy production, the cost and final price in dollars per kilowatt-hour ($/kWh).

A5.1   Physical Conditions of the Project Site

The project location is the Special Facilities Building Roof (Segment C) of the ADB headquarters in Mandaluyong City, Metro Manila, Philippines. The site is bounded by ADB Avenue on the east side, Guadix Avenue on the north side, EDSA on the west side, and a parking space on the south side.

A5.1.1   Geographical Conditions

GPS readings: WGS-84

Latitude:	14° 35' 18.91934"N
Longitude :	121° 3' 29.96689"E
Elevation:	49.69m AMSL

A5.1.2   Irradiation Readings and Meteorology

Solar radiation data are taken at the Manila Observatory, 5.55 kilometers (km) northeast of the project site and at the Philippine Atmospheric, Geophysical and Astronomical Services Administration (PAGASA) Science Garden, 6 km northeast of the project site.

---

[1]   With the exception of minor formatting and grammatical changes, this document has not been altered from the original, as provided to bidders in the ADB Rooftop Solar Power Project.

Table A5.1: Meteorological Data Summary for Science Garden in 2007

Internal beginning	Globhor kWh/m².mth	Globhor kWh/m².mth
January	116.2	69.33
February	125.8	66.18
March	170.0	76.14
April	176.2	75.60
May	171.4	74.18
June	136.0	77.85
July	144.7	83.19
August	111.8	74.06
September	127.5	70.92
October	114.1	69.23
November	100.4	65.02
December	106.8	61.39
Year	1,600.9	863.10

Globhor = global horizontal irradiance, kWh/m².mth = kilowatt-hours per meter squared per month.
Source: ADB.

A5.1.3 Available Area

Below is the roof plan of Segment C where the solar array will be installed. The roof is 151.20 meters long and 54.00 meters wide with a total area of 8,154.00 square meters.

Figure A5.1: Floor Plan of ADB Segment C Roof Deck

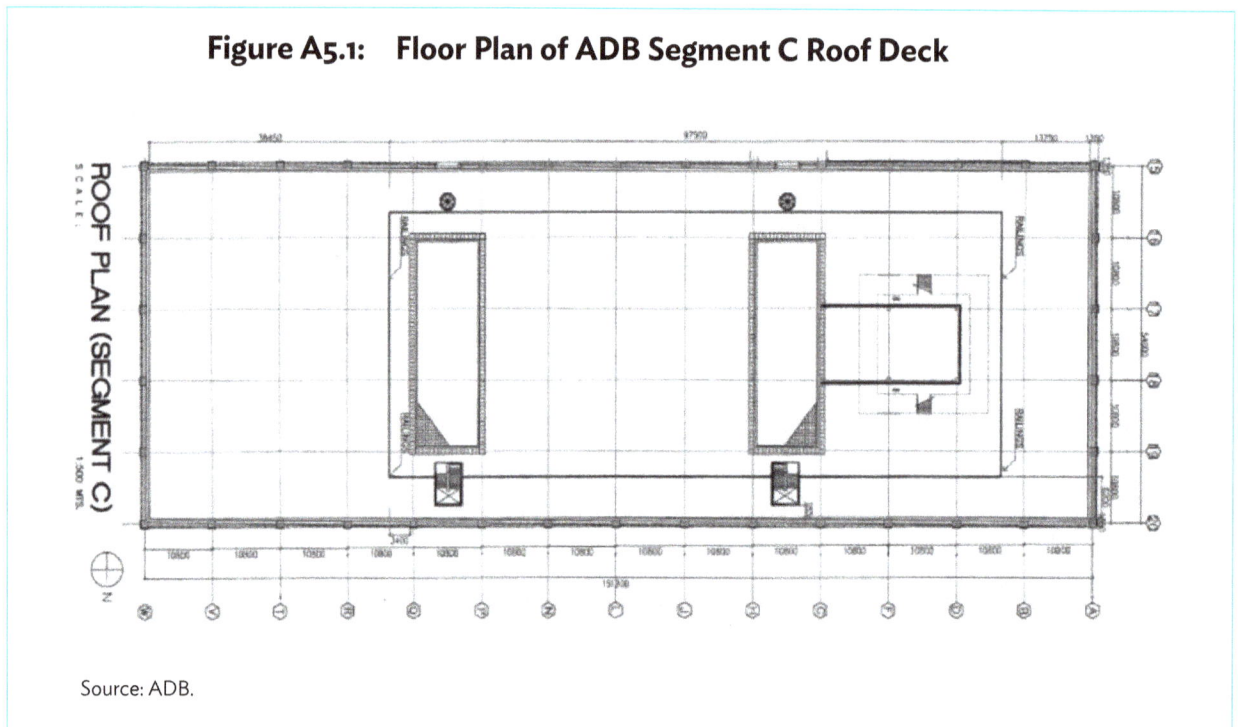

Source: ADB.

A5.1.4 Electrical Conditions

The ADB headquarters is connected to Manila Electric Company (Meralco) through two connection points. One connection is at ADB Avenue and the other is at EDSA with a contacted capacity of 4,000 kilowatts (kW) and a minimum demand of 2,800 kW. The typical load profile of ADB headquarters during regular working days and weekends are shown in Figure A5.2. The average annual energy consumption is 17 gigawatt-hours (GWh).

Figure A5.2: Average Electricity Load Profile for ADB Headquarters

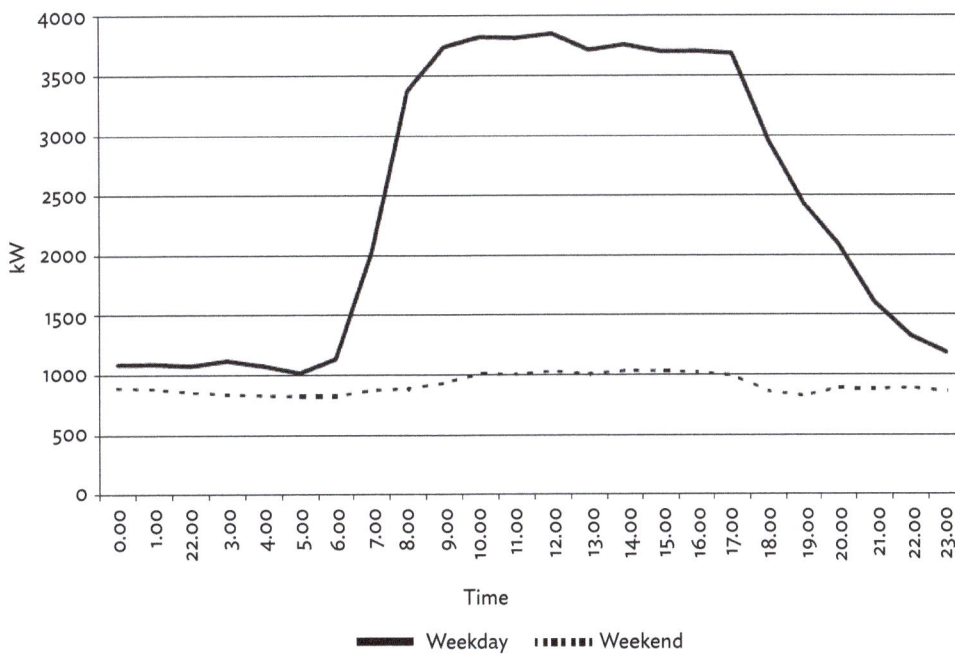

Time	Demand (kW)		Time	Demand (kW)		Time	Demand (kW)	
	Weekday	Weekend		Weekday	Weekend		Weekday	Weekend
0:00	1,089.295	892.718	8:00	3,371.553	885.105	16:00	3,704.648	1,021.245
1:00	1,092.038	883.693	9:00	3,740.220	932.558	17:00	3,688.307	993.902
2:00	1,076.514	856.245	10:00	3,824.080	1,008.614	18:00	2,951.310	861.316
3:00	1,118.410	842.009	11:00	3,817.601	1,004.639	19:00	2,425.555	821.909
4:00	1,076.128	829.122	12:00	3,851.390	1,029.778	20:00	2,093.026	888.727
5:00	1,016.195	824.470	13:00	3,716.927	1,006.717	21:00	1,607.112	875.811
6:00	1,134.684	823.350	14:00	3,759.954	1,033.419	22:00	1,328.063	889.969
7:00	2,023.721	873.691	15:00	3,701.780	1,029.672	23:00	1,182.793	858.443

kW = kilowatt.

Source: ADB.

A5.1.5   Ways of Access to Project Site

The roof area can be accessed through two staircases by personnel for inspection carrying limited equipment. Large equipment, components, and tools can be raised to the roof over two floor levels from the north side of Segment C using cranes or side construction elevators.

A5.1.6   Shading Considerations

Segment C is located in the middle of the commercial area surrounded by tall buildings. Shading on the roof of Segment C shall be considered to select the best location to mount the solar modules and to have an accurate projection of the annual production of the photovoltaic (PV) system.

A three-dimensional (3D) rendering of Segment C with the surrounding buildings is shown in Figure A5.3 using data on the dimension and location of the buildings.

Figure A5.3:   3D Rendering of Segment C and Surrounding Buildings

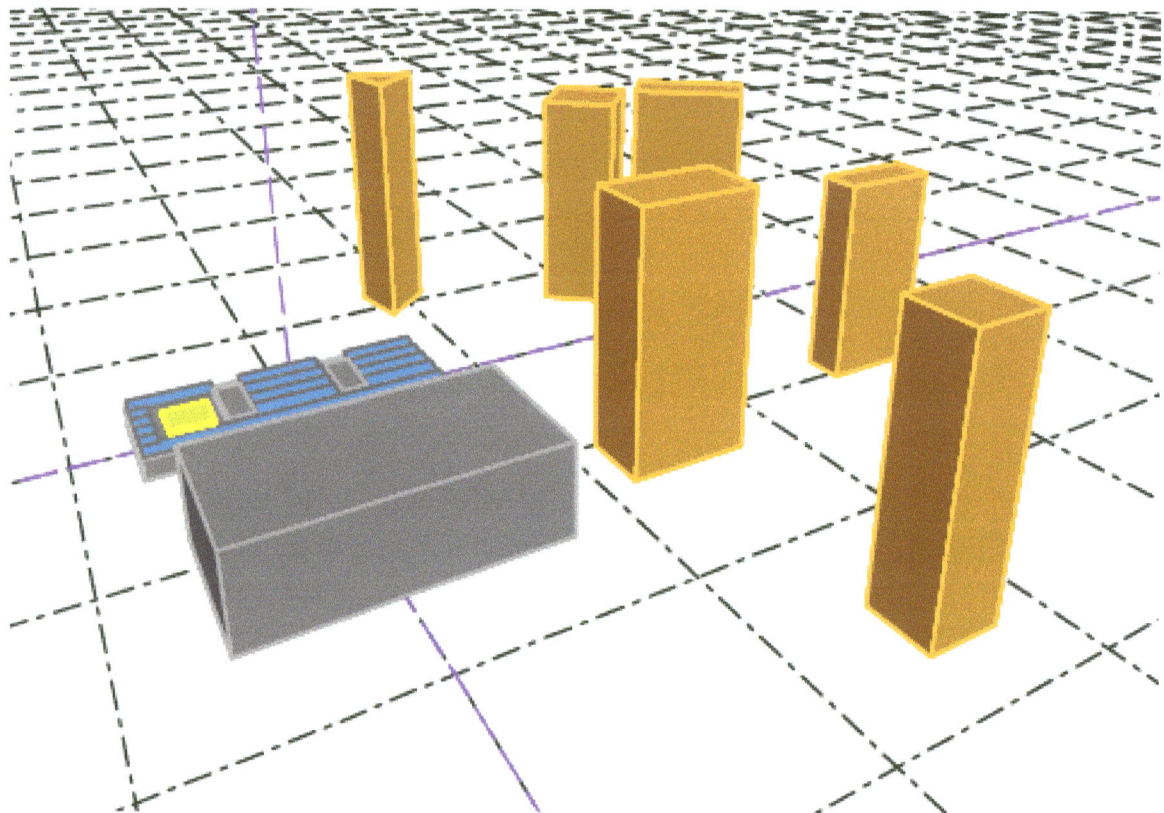

Source: ADB.

A5.1.7 Future Structures

ADB will construct additional buildings in the future that can affect the shading of the solar array on the roof of Segment C (Facilities Block). The two major structures are the Seven-Level Parking Area and the Third Atrium. These structures are added in the shading model to produce the shading map. The 3D rendering of the buildings is shown in Figure A5.4.

Figure A5.4:   3D Rendering of Segment C Including Proposed Structures

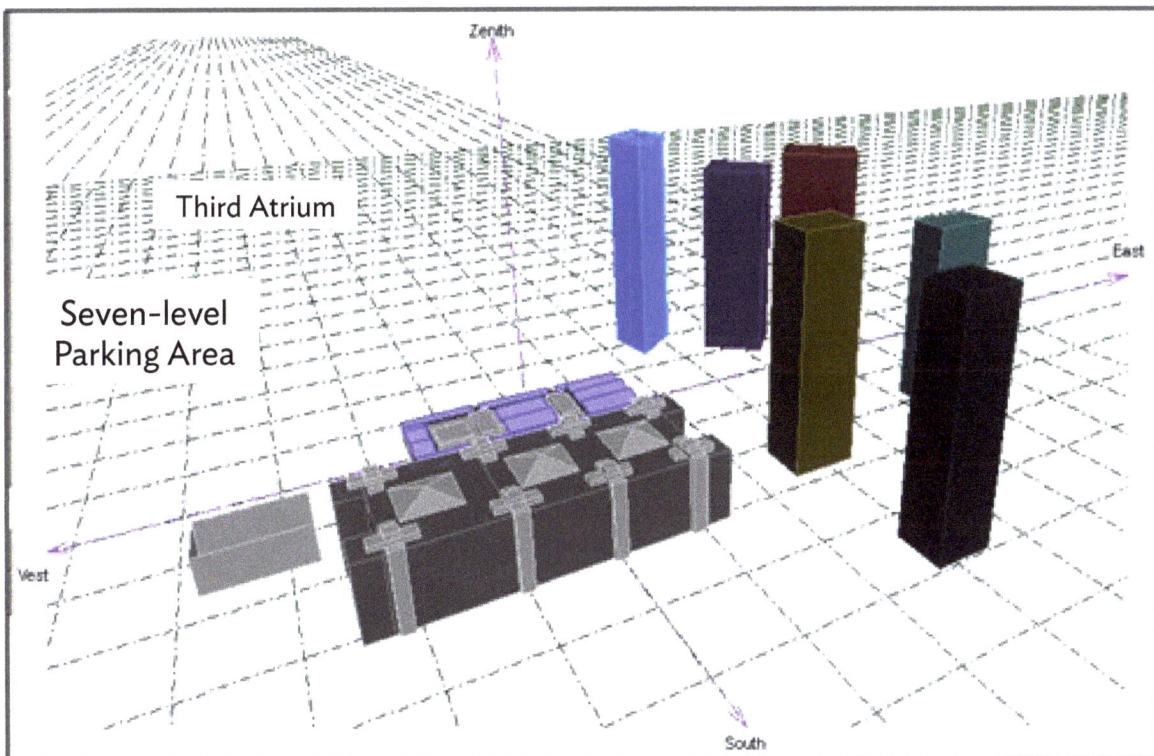

Source: ADB.

A5.2   Rooftop Solar Design

A5.2.1 Physical Sizing

The roof area is divided in sections that are usable for the solar installation. Sections such as the helipad and the roof of the fan rooms are not suitable for solar installation. The total calculated usable area is 5,166 m² and it is shown in the shaded area in Figure A5.5.

Figure A5.5: Total Usable Area Calculations

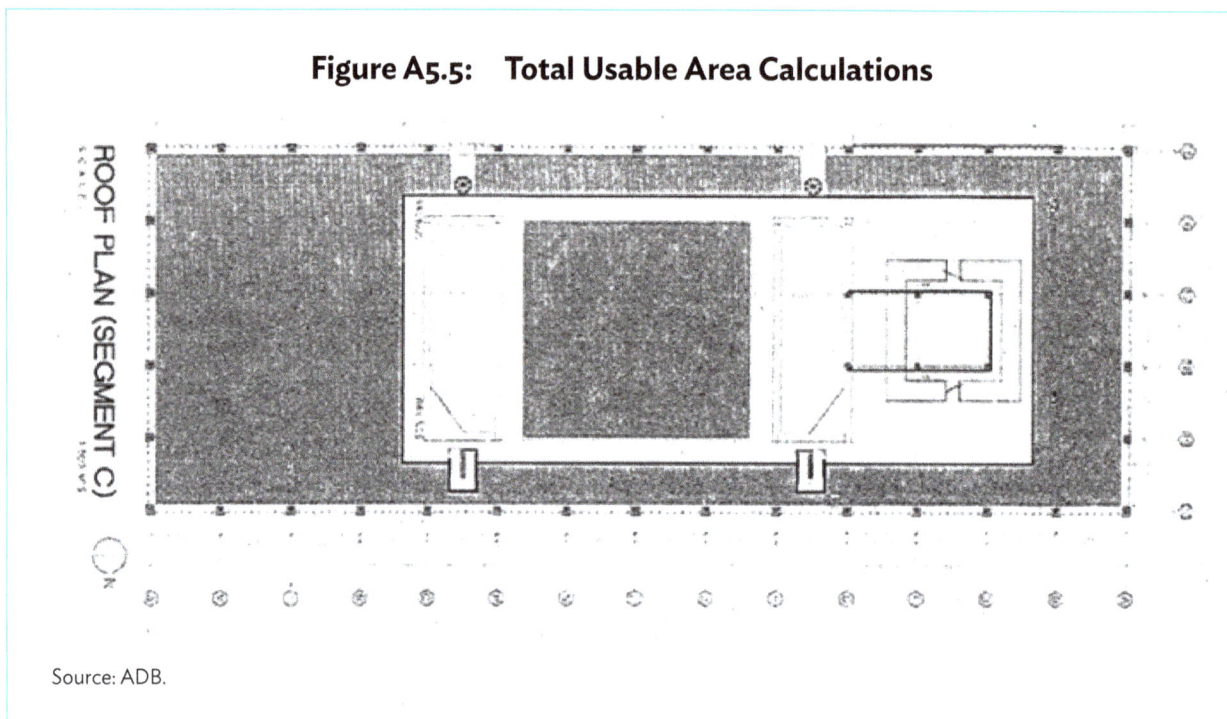

Source: ADB.

A5.2.2 Azimuth and Tilt Angles

The roof is perfectly oriented along the north to south direction. Segment C has a flat roof. Tilting the solar array toward the south naturally orients its azimuth to 0 degrees or due south.

A5.2.3 Structural Design of Support

The solar modules shall be mounted on support structures with the optimum slope and orientation to maximize its capture of the solar radiation of production of electricity. The support structures shall be ballasted without intrusions to the roof to prevent damage on the water proofing protection of the roof. Structural support is required to hold the solar modules in place with rated wind speed of 270 kilometers per hour to comply with the Philippine Electrical (PEC) Code Part 2 for Zone 1. The Civil Aviation Authority of the Philippines (PAAC) manual for aerodrome (helipads) markers, markings and signs must be constructed to withstand a wind velocity of 60 meters per second from jet blasts. The support structure must also be light enough to avoid overloading the capacity of the roof. The material used for structure must withstand corrosion from the local weather conditions with high humidity.

The Civil Aviation Authority of the Philippines (CAAP) manual for aerodrome (helipads) requires clear access to the helipad by approaching aircraft during landing and takeoff. The solar array structures around the helipad should not exceed the height of the helipad and allow access for rescue and emergency operations. To meet these requirements, the solar array structures on the west side should not exceed the height of the helipad.

The mounting frame design and array arrangement, and weights should be approved by a structural engineer for the physical integrity and reliability of the roof structures.

A5.2.4 Key Plant Components

A5.2.4.1 Solar Modules

Solar photovoltaic module performance is rated under specific conditions. The rating used is the standard test condition (STC) at cell temperature of 25°C, solar radiation of 1,000 watts per square meter (W/m²) at air mass of 1.5. Calculation of electrical characteristics shall be based upon normal operation cell temperature (NOCT) of 45°C.

Solar modules have the following specification data requirements:

Electrical Characteristics	Units
STC power rating (P_N)	Watt peak (Wp)
Peak efficiency	percentage %
Power tolerances	± 3%
Number of cells	According to manufacturer
Current at maximum power point (I_{mpp})	Amperes (A)
Voltage at maximum power point (V_{mpp})	Volts (V)
Short circuit current (I_{sc})	A
Open circuit voltage (V_{oc})	V
Maximum system voltage (V_{max})	600–1,000 V

Mechanical Characteristics	
Type	Silicon and non-silicon
Output terminal type	Multi-contact connector type 4
Length	mm
Width	mm
Depth	mm
Weight	kg

Warranty and Certifications	
80% power output warranty period	25 years

Certifications

IEC 61 215: Terrestrial PV modules with crystalline silicon
  Design qualification and type approval
IEC 61 646: Terrestrial PV modules with thin-film technology
  Design qualification and type approval

Solar modules: verification of electrical data against NOCT 45°C

The on-site performance of the solar gear should be certified by an international independent third-party certification body, i.e., Fraunhofer Institute of Germany, TÜV of Germany, Solar Institute of Singapore, Arsenal of Austria, Underwriters Laboratories Inc. (UL), etc.

A5.2.4.2 Inverters

The inverter will convert direct current (DC) electricity from the solar module and convert it to alternating current (AC) that will be synchronized with the utility grid to power the electrical loads of the facility. The type of inverters used for this application are called grid-tie inverters.

The primary requirement in considering the use of grid-tie inverters is its compliance with the UL 1741 Standard for Safety of Inverters, Converters, Controllers and Interconnection System Equipment for Use with Distributed Energy Resources as published by Underwriters Laboratories Inc. on 28 January 2010. The inverter should also be compliant with Institute of Electrical and Electronics Engineers (IEEE) 1547 standards.

Table A5.2: Inverter Specifications (Typical) – Electrical Characteristics (DC)

DC Electrical Components	Units
Maximum Input Current	A
Start Voltage	V
Maximum DC Input Voltage	V
Peak Power Tracking Voltage Range	V
Polarity of Ground	
DC Disconnect Included	
Number of String Inputs	
Fuses for String Inputs Included	
Max Rating for String Fuses	A

A = ampere, DC = direct current, V = volt.

Source: ADB.

Table A5.3: Inverter Specifications (Typical) – Electrical Characteristics (AC)

AC Electrical Components	Value (Typical)	Units
Rated Power Output		W
Compatible Utility Voltages	208 and 480	V
Minimum Efficiency	97.00	%
CEC Weighted Efficiency	97 (at 208 V)	%
Maximum Output Current		A
AC Disconnect Included	Yes	
Sleep Consumption		W

A = ampere, AC = alternating current, CEC = California Energy Commission, V = volt, W = watt.

Source: ADB.

Table A5.4: Inverter Specifications (Typical) – Mechanical Characteristics

Mechanical Characteristics	Value (Typical)	Units
Height		in (mm)
Width		in (mm)
Depth		in (mm)
Weight		lb (kg)
CSI-Approved Built-in Meter	Yes	
Cooling Method	Forced Air	

CSI = California Solar Initiative, kg = kilogram, lb = pond, mm = millimeter.

Source: ADB.

Table A5.5: Inverter Specifications (Typical) – Environmental and Certifications Compliance

Environmental Constraints		
Ambient Temperature Range	–4° to +122°/ (–20° to +50°)	°F/(°C)
Compliances		
Compliances	UL 1741, IEEE 1547	
CSI Listed	Yes	

CSI = California Solar Initiative, IEEE = Institute of Electrical and Electronics Engineers, UL = Underwriters Laboratories Inc.

Source: ADB.

Inverters: verification of electrical data against NOCT
The on-site performance of the power inverter shall be certified by an international independent third-party certification body, i.e., Fraunhofer Institute of Germany, TÜV of Germany, Solar Institute of Singapore, Arsenal of Austria, Underwriters Laboratories Inc. (UL), etc.

Inverter selection: inverter capacity and number of inverters
The capacity of the grid-tie inverter should be able to handle the total output of the solar array and convert it to utility grade electricity. The system will require multiple inverters due to the following constraints:

- **Reliability.** Multiple inverters are more reliable than a single inverter system.
- **Efficiency.** Large inverters are more efficient than smaller inverters if they are operating at their optimum capacity. The size of the inverter must be matched with the array that it will handle.
- **Solar array segmentation and shading characteristics.** Shading of a part of a string of solar modules in an array reduces the output of the whole array. Segmentation of the solar array will isolate the shaded modules from the rest of the solar array to reduce the effects of the shading. The number of inverters can be matched with the number of array groupings based on the shading characteristics of the array. The capacity of the inverter can be matched to the size of the sub array that it is assigned to handle.

- **Space constraints.** Larger inverters will occupy less total space than several small ones, but it may be that only small inverter units can fit in the available space. However, the space available may only fit small inverter units instead of a few large ones. Large inverters may also require special equipment such as forklifts, cranes, or chain blocks, which the space or the structure may not be able to accommodate.
- **Environmental conditions.** Inverters should ideally be housed in power control rooms together with the rest of the switch gears of the facility. Outdoor units may be practical if there is a limitation in the availability of indoor space. Extra protective coating for the inverters is required for corrosive environments.

Input Voltage and Capacity Considerations

The selection of the right inverter for the system depends on the solar array configuration, particularly the operating voltage, the operating current, and the maximum open circuit voltage of the array. As illustrated by Figure A5.6, the solar array has an optimum operating voltage at which the inverter must function with the highest conversion efficiency at its rated power. The solar array will have a lower operating voltage at low solar radiation levels especially during early morning, late afternoon, and during cloudy days. The inverter must be able to start its operation at lower than the array operating voltage (V_{MPPT}) to produce output for longer periods of time. The solar array will produce the highest voltage when it is not loaded, or in open-circuit mode, and when the temperature is cold. The inverter input must be able to handle this open-circuit voltage without damage.

Figure A5.6: Solar Array Voltage Characteristics

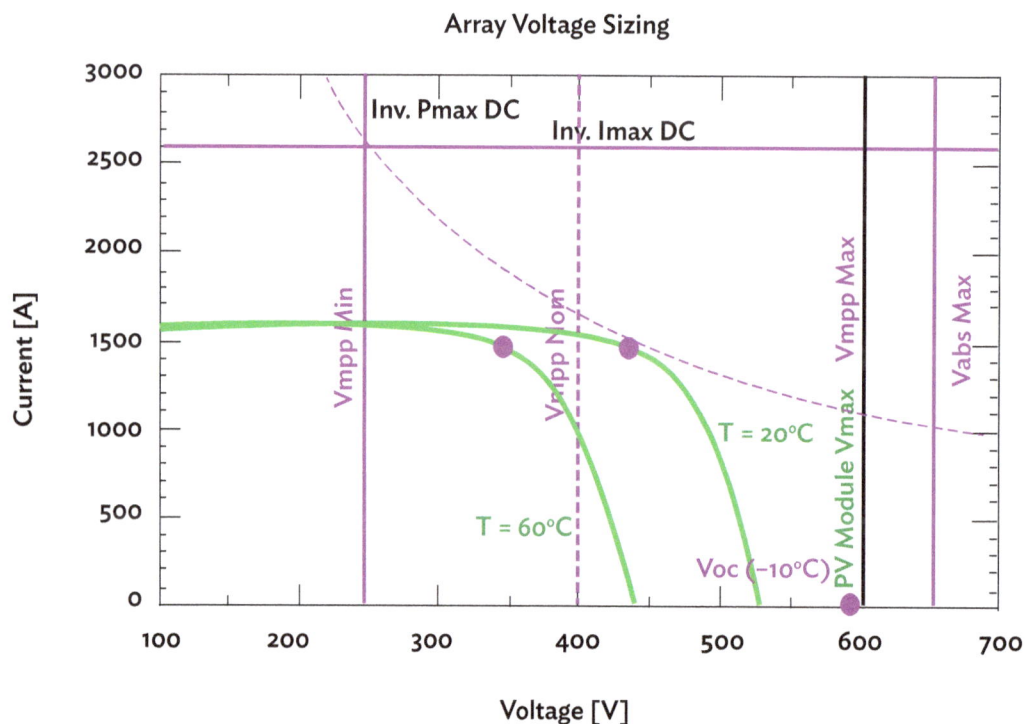

Inv. Imax DC = Maximum inverter direct current, Inv. Pmax DC = DC voltage at maximum inverter power, mpp = maximum power point, PV Module Vmax = Maximum system voltage specified for the PV Module, T = PV module operating temperature, Vabs Max = Absolute maximum inverter input voltage, Vmpp = voltage at the point where the most power is delivered, Vmpp Max = Maximum inverter operating voltage, Vmpp Min = Minimum inverter operating voltage, Vmpp Norm = Normal inverter operating voltage, Voc (–10°C) = Maximum array absolute voltage.

Source: ADB.

Output Parameter Considerations

The inverter output must match the interconnection requirement of the grid or the internal power system. In the case of ADB, the most ideal connection will be at the 34.5 kilovolt (kV) level, three-phase output, with neutral and ground wires, with a frequency of 60 hertz (Hz). Considering that the interconnection will be done at substation 1 where only 34.5 kW and 380 V Y/220 V voltage levels are available, tapping at the low voltage level will be practical. The inverter shall have a three-phase delta connection with 230 V line-to-line, or a three-phase wye output with 380 V line to line with a common neutral.

A5.2.4.3   Electrical Wiring and Interconnections

The system shall be compliant with the Philippine Electrical Code (PEC) under the following provisions:

PEC 6.90.1.5   *Roof mounted DC photovoltaic arrays located in dwellings shall be provided with DC ground fault protection to reduce fire hazards.*

PEC 6.90.1.4 (b)   *Conductors of different systems shall not be contained in the same raceways, cable tray, cable, outlet box, junction box, or similar fitting as feeders of branch circuits of other systems, unless the conductors of the different systems are separated by a partition or are connected together.*

PEC 6.90.2.2 (4.b.1)   *The circuit conductors and over current devices shall be sized to carry not less than 125% of the maximum currents.*

PEC 6.90.4   *Wiring method for solar PV cells:*

(b) Single conductor cable. Type SE, UF, USE, and USE-2 single-conductor cable shall be permitted in photovoltaic source circuits where installed in the same manner as a Type UP multiconductor cable in accordance with Article 3.39. Where exposed to direct rays of the sun, Type UP cable identified as sunlight-resistant or Type USE cable shall be used.

(c) Flexible Cord and Cables. Flexible cords and cables where used to connect the moving parts of tracking PV modules, shall comply with Article 4.0 and shall be of a type identified as a hard service cord or portable power cable; shall be suitable for extra-hard usage, listed for outdoor use, water resistant, and sunlight resistant.

(d) Small Conductor Cables. Single-conductor cables listed for outdoor use that are sunlight resistant and moisture resistant with sizes 1.25 mm² (1.2 mm dia.) and 0.75 mm² (1.0 mm dia.) shall be permitted for module interconnections where such cables meet the ampacity requirement of 6.90.2.2 section 3.10.1.15 shall be used to determine the cable ampacity and temperature derating factors.

(e) Direct-Current Photovoltaic Source and Output Circuits inside a Building. Where direct current photovoltaic source or output circuit of a utility-interactive inverter from a building-integrated or other photovoltaic system are run inside a building or structure, they shall be contained in metallic raceways or enclosures from the point of penetration of the surface of the building or structure to the first readily accessible disconnecting means. The disconnecting means shall comply with 6.90.3.2(a) through 6.90.3.2(d).

PEC 6.90.4.2   *Component Interconnections.* Fitting and connectors that are intended to be concealed at the time of on-site assembly, where listed for such use, shall be permitted for on-site interconnection of modules or other array components. Such fittings and connectors shall be equal to the wiring method employed in insulation, temperature rise, and fault-current withstand, and shall be capable of resisting the effects of the environment in which they are used.

PEC 6.90.4.3 *Connectors. The connectors permitted by Article 6.90 shall comply with PEC 6.90.4.3(a) through 6.90.4.3(e):*

(i)   Configuration. The connectors shall be polarized and shall have a configuration that is non-interchangeable with receptacles in other electrical system on the premises.

(ii)   Guarding. The connectors shall be constructed and installed so as to guard against inadvertent contact with live parts by persons.

(iii)   Type. The connector shall be of the latching or locking type.

(iv)   Grounding Member. The grounding member shall be the first to make and the last to break contact with the mating connector.

(v)   Interruption of Circuit. The connectors shall be capable of interrupting the circuit without hazard to the operator.

PEC 6.90.5   *Grounding*

PEC 6.90.5.1   *System Grounding. For a photovoltaic power source, one conductor of a 2-wire system rated over 50 volts and a neutral conductor of a 3-wire system shall be solidly grounded. Exception: Systems complying with 6.90.4.5. (Ungrounded PV systems)*

Summary of Required Protection

Overcurrent protection. All current carrying conductors shall be equipped with overcurrent protection rated at 125% of its nominal current capacity.

Disconnecting means (isolation switch). The system shall have disconnecting means or isolation switch from the module outputs and inverter outputs for maintenance and emergency purposes.

Grounding (earthing). Non-current carrying metal parts of the system, such as module mounting frame and inverter chassis, shall be wired and bonded to ground to direct stray currents.

Lightning protection. Air terminals shall be installed to place the solar array within its zone of protection. All connections shall be bonded to have the lowest resistance and connected to the earthing rod.

Anti-islanding protection. The grid-interactive solar photovoltaic system shall automatically de-energize its output to the connected electrical production and distribution network upon loss of voltage in that system and shall remain in that state until the electrical production and distribution network voltage has been restored.

Inverter Monitoring Features

The inverter shall have a communication interface that will provide the user a remote monitoring, data logging, and control of the system. The following parameters shall be monitored:

- Grid side (3 phases): AC voltage, current, frequency, and power factor
- Inverter output (3 phases): AC voltage, current, frequency, and power factor
- Solar array side (per major string): DC voltage, current
- Operational: Ground fault, temperatures, relative humidity, solar radiation, wind velocity, and direction

Tie-in to substation

Interconnection of the solar system will be done at substation 6 at 230 VΔ/380 VY tapping voltage as shown in Figure A5.7. The inverter can have a three-phase delta connection with 230 V line-to-line, or a three-phase wye output with 380 V line to line with a common neutral. The physical connection will be located inside the substation cabinet tapped into the bus bar. The substation is located at the ground level of segment C on the southeastern side together with substation 7. This shall be compliant with the PEC 2009 section 6.90.7.

Figure A5.7: Site Development Plan and Interconnections

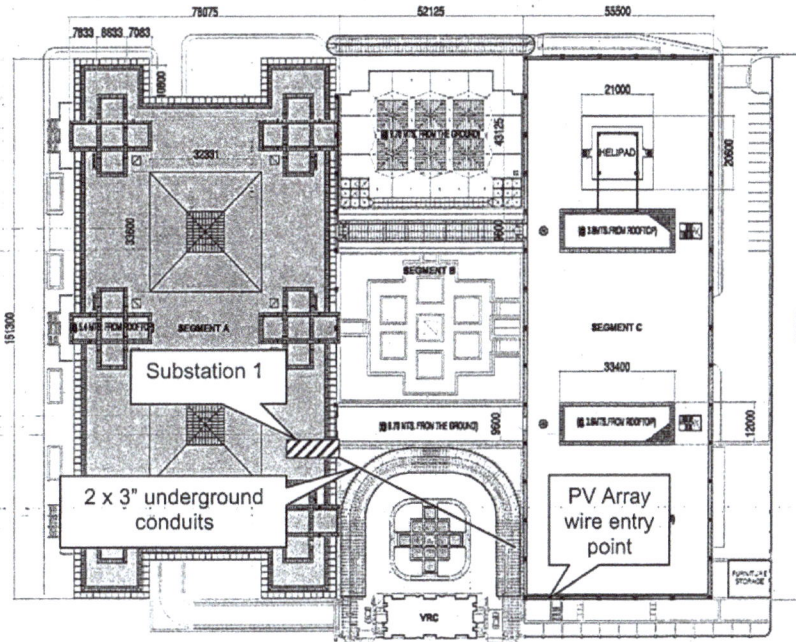

Substation 1 Details *

Substation #1			
ADB Side (3MVA)		EDSA Side (3MVA)	
Demand	310kW	Demand	350kW
Loads:		Loads:	
Data Center UPS (Normal)		Data Center ACU (Normal)	
Capacitor Bank		Core 6 Elevators	
Mechanical Bus Duct (AHU's, FPUs, etc.)		Mechanical Bus Duct (AHU's, FPUs, etc.)	
Lighting & Receptacle Bus Duct (Lights & C.O.)		Lighting & Receptacle Bus Duct (Lights & C.O.) UPS 6	
1 Spare		1 Spare	

Substation 1 MV Switch gear (Front view) typical

Metering	Bus Duct A4 Lights and Receptacles	Bus Duct A3 Mechanical	34.5kV to 380Y/220V Substation	Bus Duct A2 Mechanical	Bus Duct A1 Lights and Receptacles	Metering
	12 ATS -1 A5	Capacitor Bank	1 Core No.6	Spare**	32 EATS -1 A7	
Main Breaker 1-2	Spare**	Full Meter	Tie Breaker	Full Meter	HO ATS -1 HO DP - 1	Main Breaker 1-1
30"	22"	22"	30"	22"	22"	30"

* The bidders are free to evaluate personally suitability of other substations for the solar interconnection.
** Requires 2 pieces of RL1600 circuit breakers if used for interconnection

Source: ADB.

91

A5.2.5 Rooftop Solar Performance

The list of losses of the individual component and the total system loss shall be as illustrated in Figure A5.8.

Calculations of individual losses:

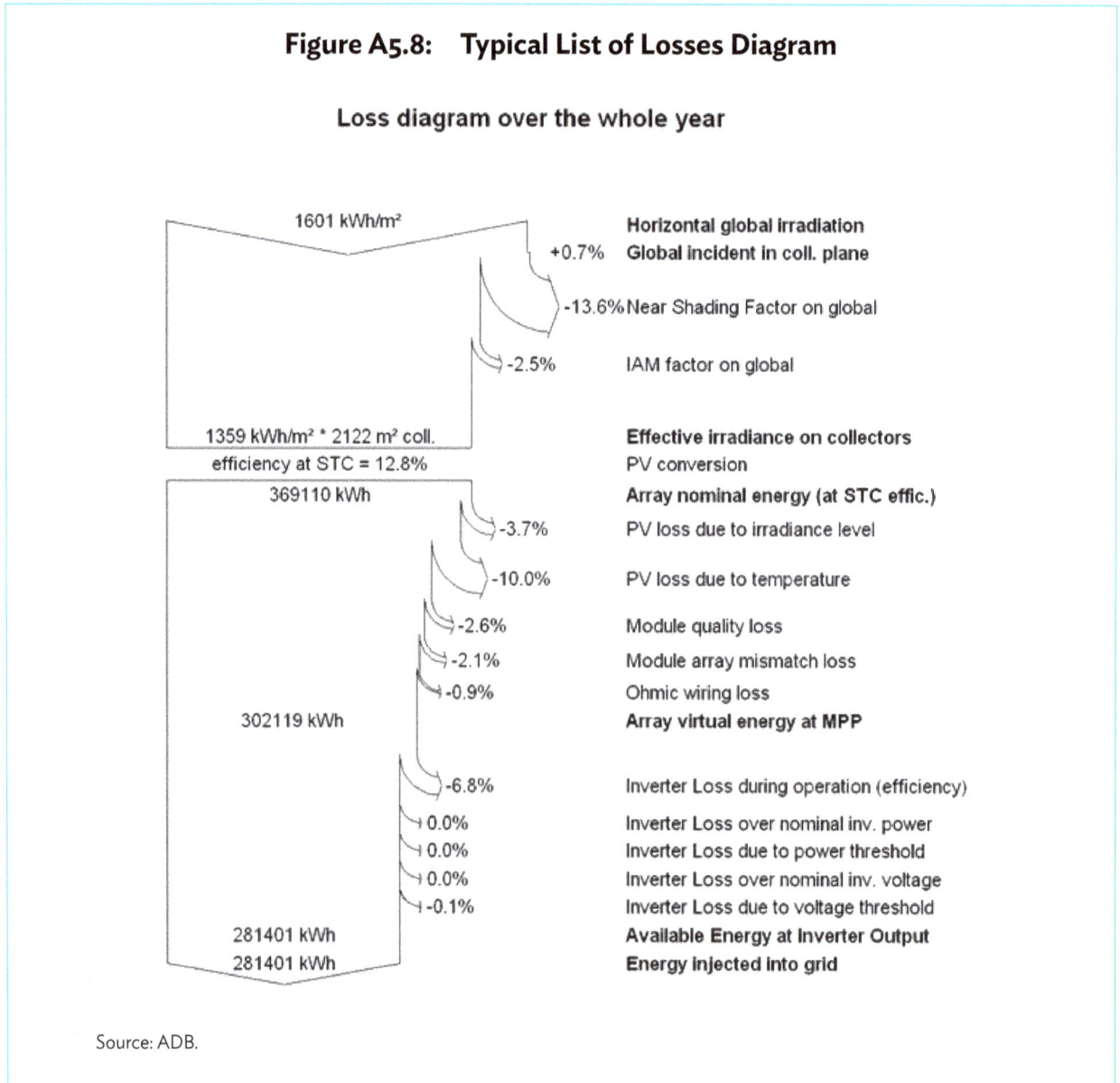

Figure A5.8: Typical List of Losses Diagram

Loss diagram over the whole year

1601 kWh/m²	**Horizontal global irradiation**
+0.7%	Global incident in coll. plane
-13.6%	Near Shading Factor on global
-2.5%	IAM factor on global
1359 kWh/m² * 2122 m² coll. efficiency at STC = 12.8%	**Effective irradiance on collectors** PV conversion
369110 kWh	**Array nominal energy (at STC effic.)**
-3.7%	PV loss due to irradiance level
-10.0%	PV loss due to temperature
-2.6%	Module quality loss
-2.1%	Module array mismatch loss
-0.9%	Ohmic wiring loss
302119 kWh	**Array virtual energy at MPP**
-6.8%	Inverter Loss during operation (efficiency)
0.0%	Inverter Loss over nominal inv. power
0.0%	Inverter Loss due to power threshold
0.0%	Inverter Loss over nominal inv. voltage
-0.1%	Inverter Loss due to voltage threshold
281401 kWh	**Available Energy at Inverter Output**
281401 kWh	**Energy injected into grid**

Source: ADB.

ITEM:	Rating:
Horizontal global irradiation	kWh/m²
Global incident on collector plane	+%
Near shading factor on global	−1%
IAM factor in global	−%
Effective irradiance on collectors	kWh/m² * m² collector
PV conversion	Array efficiency at STC = %
Array nominal energy (at STC efficiency)	369,110 kWh
PV loss due to irradiance level	−%
PV loss due to temperature	−%
Module quality loss	−%
Module array mismatch loss	−%
Ohmic wiring loss	−%
Array virtual energy at MPP	kWh
Inverter losses	
Inverter loss during operation (efficiency)	−%
Inverter loss over nominal inverter power	−%
Inverter loss due to power threshold	−%
Inverter loss over nominal inverter voltage	−%
Inverter loss due to voltage threshold	−%
Available energy at inverter output	kWh
Energy injected to grid	kWh

A5.2.6 Technofinancial Model

The bidder shall present its own technofinancial model to evaluate the yearly energy production volume, the cost and final price of energy (in $/kWh) of the rooftop solar project. The technofinancial model shall consider the system capacity in kilowatts-peak with the proposed capital cost in US dollars. Provisions of loans from ADB and Private Sector Operations Department (PSOD) to augment and provide financial requirement of the service provider to cover the capital cost of the system shall be considered. Terms and conditions of the financial mix are part of the tender documents.

Assumptions

The financial runs using the technofinancial model shall show the following assumptions based on industry practice and conservative estimates:

Solar modules	$/Wp
Mounting structures	$/Wp
Inverter	$/Wp
Wiring and interconnection:	$/Wp
Labor for installation	$/Wp
Others (permitting, etc.)	$/Wp
Construction time insurance	%
Contingency	%
Annual O&M and insurance (% of project cost)	%
Average daily solar radiation (PAGASA 2007 Data)	kWh/m²/day
Plant capacity (STC rating)	kW

Array area/kWp (module technology-specific)	m²/kWp
Total array area (shall not exceed 6,400 m²)	m²
Shading factor (+albedo)	%
Efficiencies:	
PV array conversion efficiency	%
PV annual production degradation	%
Inverter efficiency	%
Wires and cable efficiency	%
PV efficiency due to heat	%
Availability (due to downtime, grid outages, etc.)	%
Average load factor (utilization)	%

Project Performance

The technofinancial study shall show the projected cost of energy (in $/kWh) on a fixed or flexible cost within the established project period of 15 years. The payback of the system shall be shown as well as the annual displacement of carbon dioxide (in tons per year) using the assumed emission factor of 0.5 kg/kWh for the Luzon grid.

Annexes to Project Technical Outline (not included here):

- Mandaluyong City Requirements for Building Permit Application
- Building Permit Form (Mandaluyong City)
- Electrical Permit Form (Mandaluyong City)
- Mechanical Permit Form (Mandaluyong City)
- Barangay Wack-Wack Map
- CAAP Permit to Operate
- Photographs of Helipad Improvements
- Relevant Sections of the Philippine Electrical Code (2009)
- Relevant Sections of the CAAP Manual for Aerodromes
- Relevant Sections of the Philippine Distribution Code of 2001
- Philippine Renewable Energy Law of 2008 (RA 9513)
- Implementing Rules and Regulations for RE Law RA 9513
- PAGASA Global Radiation Data 2007–2008
- Spherical Photographs from the Roof with Solar Path Diagram

ANNEX 6
Sample Solar Services and Site Lease Agreement

Solar Services and Site Lease Agreement

(Contract No. _____)

This Solar Services and Site Lease Agreement (this "Agreement") made on the _____day of __ [*month*]__, 20_ _ between:

[*Name of the Power Purchaser*], [*description of organization type and legal status*], with offices at [*physical address of offices*] (hereinafter called the "Power Purchaser"), represented herein by [*name, job title and division of power purchaser representative*];

-and-

[*Name of the Power Provider*], a corporation incorporated under the laws of [*country of the Power Provider*] and having its principal place of business at [*address of the Power Provider*] (hereinafter called the "Power Provider"), represented herein by [*name, job title and division of power provider representative*];

WHEREAS, the Power Purchaser is promoting the development and utilization of renewable energy sources;

WHEREAS, the Power Purchaser desires to purchase from the Power Provider and the Power Provider desires to sell to the Power Purchaser the entire energy output of the Generating Facility;

WHEREAS, the Power Purchaser as Lessor, is the owner of the Site;

WHEREAS, the Power Provider desires to install an electricity photovoltaic, solar power plant with a total generating capacity rated at approximately _____ kWp (referred to as the "Generating Facility") located on the rooftop of the [*building name, organization name, and physical address*] (the "Site");

WHEREAS, the Power Provider desires to lease certain premises from the Power Purchaser in order to install, maintain and operate the Generating Facility, and the Power Purchaser desires to permit such installation, maintenance and operation on the terms and conditions herein contained;

WHEREAS, following a competitive selection process, the Power Purchaser has accepted a bid by the Power Provider for the provision of solar power to the Power Purchaser;

NOW, THEREFORE the Power Purchaser and the Power Provider agree as follows:

1. CONTRACT DOCUMENTS - ORDER OF PRECEDENCE – DEFINITIONS

1.1 Contract Documents

The following documents shall constitute the Contract between the Power Purchaser and the Power Provider, and each shall be read and construed as an integral part of the Contract:

(a) This Agreement and the Appendices hereto;

(b) Letter of Price Bid and Grand Price Summary submitted by the Power Provider;

(c) Letter of Technical Bid and Technical Proposal submitted by the Power Provider;

(d) [*Name of project*] Technical Outline;

(e) Other completed Bidding Forms submitted with the Letters of Technical and Price Bid;

1.1 Order of Precedence

In the event of any ambiguity or conflict between the Contract Documents listed above, the order of precedence shall be the order in which the Contract Documents are listed in Section 1.1 (Contract Documents) above.

1.1 Appendices

The Appendices listed in the attached List of Appendices shall be deemed to form an integral part of this Agreement.

Reference in this Agreement to any Appendix shall mean the Appendices attached hereto, and this Agreement shall be read and construed accordingly.

1.1 Definitions

(a)   "Access Property" has the meaning given in Section 2.1.

(b)   "Authorized Representatives" are the persons named in Section 23 below and authorized to represent the Parties in sending and/or receiving Notices and Communications related to this Agreement.

(c)   "Contract Documents" consist of this Agreement and its Appendices, including but not limited to, the Technical Specifications, Technical and Financial Bid Proposals, Letter of Acceptance of Bid, the Power Purchaser's Fire, Safety and Security Regulations, relevant Administrative Orders, Good Social Management Certificate, and amendments thereto.

(d)   "Corrupt, Fraudulent, Coercive, and Collusive Practices" refer to acts or omissions prohibited under the Power Purchaser's Anti-Corruption Policy, as discussed in Section 34 below.

(e)   "Default" means any breach of the obligations of a Party (including but not limited to breach of a fundamental term) or any other default, act, omission, negligence or negligent statement of a Party in connection with or in relation to the subject matter of the Contract and in respect of which such Party is liable to the other.

(f)   "Delay" means delivery beyond the Delivery Date or Dates specified in this Agreement.

(g)   "Energy Delivery Point" means the energy delivery point within the Site's electrical system on the Power Purchaser Site's Utility meter.

(h)   "Energy Output" means the total quantity of all actual net energy generated by the Generating Facility (measured in kWh AC) and delivered in accordance with Section 9 hereof to the Energy Delivery Point, in any given period of time. Energy Output does not include the Environmental Incentives.

(i)   "Environmental Attributes" means the characteristics of electric power generation at the Generating Facility that have intrinsic value, separate and apart from the Energy Output, arising from the perceived environmental benefits of the Generating Facility, including but not limited to all environmental and other attributes that differentiate the Energy Output from energy generated by fossil-fuel based generation units, fuels or resources, characteristics of the Generating Facility that may result in the avoidance of environmental impacts on air, soil or water, such as the absence of emission of any oxides of nitrogen, sulfur or carbon or of mercury, or other gas or chemical, soot, particulate matter, or other substances attributable to the Generating Facility or the compliance of the Generating Facility or the Energy Output with the law, with the rules and standards of the United Nations Framework Convention on Climate Change (the "UN-FCCC") or with the Kyoto Protocol to the UNFCCC or crediting "early action" with a view thereto, or laws or applicable counterpart program in [*name of country*].

(j)   "Environmental Incentives" mean all rights, credits (including tax credits), rebates, benefits, reductions, offsets, and allowances and entitlements of any kind, howsoever entitled or named (including carbon credits and allowances), whether arising under national or local law, international treaty, trade association membership or the like arising from the Environmental Attributes of the Generating Facility or the Energy Output or otherwise from the development or installation of the Generating Facility or the production, sale, purchase, consumption or use of the Energy Output. Without limiting the forgoing, "Environmental Incentives" include green tags, renewable energy credits, tradable renewable certificates, portfolio energy credits, the right to apply for (and entitlement to receive) incentives under the [*names of renewable energy laws and regulations, e.g. Implementing Rules and Regulations of Republic Act 9513 or the Renewable Energy Act of 2008 of the Philippines*].

(k)   "Force Majeure" or "Fortuitous Event" refers to events, which the Power Provider could not have foreseen, or which though foreseen, were inevitable, as further defined in Section 20 below. It shall not include ordinary unfavorable weather conditions and any other cause the effects of which could have been avoided with the exercise of reasonable diligence by the Power Provider.

(l)   "Generating Facility" means the electric power generation equipment, controls, meters, switches, connections, conduit, wires, and other equipment connected to the Energy Delivery Point, installed by the Power Provider as a fixture on the Site for the purpose of providing electric power to the Power Purchaser under this Agreement.

(m)   "Kiosk" means a small stand-alone device providing live PV system related information via visual display which will be located on the Site.

(n)   "kW AC" means kilowatt alternating current.

(o)   "kWh AC" means kilowatt-hour alternating current.

(p)   "kWp" means kilowatt peak rated power.

(q) "Notices" – refers to all written communication required under this Agreement to be exchanged between the Parties, including but not limited to, requests, permissions, or consent.

(r) "Operation Date" means the date on which the Generating Facility has achieved Operation.

(s) "Operation" means an event that is deemed to occur when the Generating Facility is (i) mechanically complete and operating and (ii) energy is delivered through the Generating Facility's meter and to the Site's designated electrical system.

(t) "Party" or "Parties" refer to either or both the Power Purchaser and the Power Provider.

(u) "Permits" means all governmental permits, licenses, certificates, approvals, variances, and other entitlements for use necessary for the installation and Operation of the Generating Facility.

(v) "Premises" means that portion of the Site on which the Generating Facility shall be constructed and installed.

(w) "Project" means the photovoltaic rooftop project consisting of support structure and the Generating Facility to be installed by The Power Provider.

(x) "Property" means the Premises and Access Property collectively.

(y) "Reporting Rights" means the right of the Power Provider to report to any national or local agency, authority or other party, including without limitation under [applicable country energy laws] or under any present or future domestic, international or foreign emissions trading program, that the Power Provider owns the Environmental Attributes and the Environmental Incentives associated with the Energy Output.

(z) "Site" means the Power Purchaser's facility at [physical address].

(aa) "Term" has the meaning given in Section 6.

(ab) "Utility" means the electric distribution company responsible for electric energy transmission and distribution service at the Site.

2. LEASE

2.1 **Lease.** The Power Purchaser does hereby lease to the Power Provider in accordance with the terms and conditions hereinafter set forth, the [total installable roof area] square meters of the real property located on the rooftop of the Site and described in further detail attached hereto as Exhibit A (individually and collectively, the "Premises") for the sole purposes of installing, maintaining and operating the Generating Facility. The Power Purchaser hereby also grants to the Power Provider, for the Term as defined in Section 6, a non-exclusive right-of-way for vehicular and pedestrian ingress and egress to the Premises or the Generating Facility to the extent required by the Power Provider and as mutually agreed upon by the Parties (the "Access Property"). The Premises and Access Property shall collectively be referred to as the "Property."

2.2 **Benefits.** The Power Provider hereby covenants to pay the Power Purchaser, on or before the Operation Date, as and for rent of the Premises for the duration of the Term as defined in Section 6.3, the sum of **USD _[lease amount in Arabic numerals_]([_lease amount in words_] US Dollar)** per square meter.

3. ACCESS TO PREMISES – SECURITY AND SAFETY – EQUIPMENT

3.1 Access

The Power Provider will give the Power Purchaser reasonable written or telephonic notice before any entry onto the Premises by the Power Provider's employees, agents or contractors. The Power Purchaser will make available to the Power Provider access to the Generating Facility and the Premises for the purposes set forth in Section 4. Notwithstanding anything to the contrary in this Agreement, the Power Provider shall be permitted to access the Premises twenty-four (24) hours a day, seven (7) days a week for emergency purposes, as reasonably determined by the Power Provider. Within twenty-four (24) hours of such emergency access, the Power Provider shall provide the Power Purchaser with a written explanation of the nature of the emergency. All such emergency work shall be diligently prosecuted to completion to the end that such work shall not remain in a partly finished condition any longer than necessary for completion.

3.2 Security, Safety Rules, and Regulations

(a) The Power Provider hereby undertakes to comply with and observe the requirements of the Safety and Security Procedures (see **Appendix 5)** issued by the Power Purchaser, a copy of which has been provided to the Power Provider. Said Power Purchaser Safety and Security Procedures, as amended from time to time, shall be regarded as having been incorporated herein as an integral part of this Agreement. The Power Provider shall ensure that all its personnel who shall be engaged in the performance of the Services shall become familiar with and shall comply with and observe the rules and procedures therein set forth.

(b) The Power Provider shall take all necessary precautions for the safety of those employed on the installation/Operation activities on the Site and shall comply with all applicable provisions of national or municipal safety laws to prevent accidents or injury to persons on, about or adjacent to the Power

Purchaser's Premises. The Power Provider shall, as required by the conditions and progress of the Services, properly maintain at all times all necessary safeguards for the protection of its personnel, the Power Purchaser's employees and the public.

3.1 Tools, Equipment, Materials, and Supplies

(a)   All tools, equipment, materials, and supplies necessary to perform the work in accordance with the terms of this Agreement shall be provided by the Power Provider at no cost to the Power Purchaser. Tools and equipment brought by the Power Provider into the Power Purchaser's premises shall be checked by the Power Purchaser's Security Guard prior to allowing entry and should be appropriately covered by gate passes to be issued by the Power Purchaser's authorized representative when the tools and equipment are taken out of the Power Purchaser's Premises.

(b)   The Power Provider shall maintain in its custody supplies and materials needed for carrying out the required Operation activities in an amount adequate for the uninterrupted performance of the Operation.

4.   INSTALLATION, OPERATION, AND OWNERSHIP OF THE GENERATING FACILITY

(a)   The Power Purchaser hereby consents to the installation of the Generating Facility on the Premises, including, without limitation, solar panels, mounting substrates or supports, wiring and connections, power inverters, service equipment, metering equipment, and other interconnections.

(b)   The Power Provider shall (i) use commercially reasonable efforts to complete the installation of the Generating Facility and to begin Operation of the Generating Facility on or before *[agreed date]*; provided, however, in the event that the necessary financing, permits, authorities and agreements contemplated in parts (i) – (iv) of Section 4(c) are not completed by *[agreed date]*, the Parties may mutually agree to amend this Agreement to revise the Operation Date.

(c)   The Power Provider shall be responsible for all costs and the performance of all tasks required for installation of the Generating Facility. By no later than *[agreed date]*, the Power Provider shall commence pre-installation activities relating to the Generating Facility, which shall include, without limitation, using commercially reasonable efforts to:

(i)   obtain (A) funding for construction of the Generating Facility; and (B) self-generation incentive credits for operation of Generating Facility;

(ii)   obtain all permits, contracts, and agreements required for the installation of the Generating Facility;

(iii) *[OPTION]* effect the execution of all agreements required for Utility interconnection of the Generating Facility;

(iv) enter into contract(s) for installation of the Generating Facility, subject to the terms of the proposed financing.

(d) Unless otherwise agreed with the Power Purchaser, the Power Provider, at no additional cost or expense to the Power Purchaser, will obtain all Permits necessary for the installation and Operation of the Generating Facility. The Power Purchaser hereby gives its consent to any action taken by the Power Provider in applying for any and all Permits the Power Provider finds necessary or desirable for the operation of the Generating Facility, and the Power Purchaser hereby appoints the Power Provider its agent for applying for such Permits. The Power Provider will carry out the activities set forth in this Section 4 in accordance with all applicable laws, rules, codes and ordinances and in such a manner as will not unreasonably interfere with the Power Purchaser's operation or maintenance of the Site.

(e) Title to the Generating Facility and all improvements placed on the Premises by the Power Provider shall be held by the Power Provider during the Term. At the expiration of the Term, ownership of the Generating Facility and all improvements placed on the Premises by the Power Provider shall revert to the Power Purchaser at no charge to the Power Purchaser. The Power Provider shall take whatever actions necessary to transfer fee title ownership of the Generating Facility and all improvements placed by the Power Provider on the Premises to the Power Purchaser, free and clear from any lien or monetary encumbrance. Should this Agreement be terminated for any reason prior to the expiration of the Term, other than as provided in Section 4(b) or as a result of the Power Purchaser exercising its option to purchase the Generating Facility, as provided for in Sections 6.1 (ii) and 6.3 (c)(ii), all of the Generating Facility and the improvements placed by the Power Provider on the Premises shall remain the property of the Power Provider and shall be removed by the Power Provider within *[agreed period of time]* upon termination of this Agreement.

(f) Except as otherwise expressly provided in Section 4(g), the Power Purchaser acknowledges and agrees that notwithstanding that the Generating Facility is a fixture on the Premises, the Power Purchaser has no ownership interest in the Generating Facility and the Power Provider is the exclusive owner and operator of the Generating Facility, that the Generating Facility may not be sold, leased, assigned, mortgaged, pledged or otherwise alienated or encumbered (collectively, a "Transfer") with the fee interest or leasehold rights to the Property by the Power Purchaser. The Power Purchaser shall give the Power Provider at least *[agreed period of time]* written notice prior to any transfer of all or a portion of the Property identifying the transferee, the portion of Property to be transferred and the proposed date of transfer. The

Power Purchaser agrees that this Agreement and the right-of-way granted in Section 3.1 shall run with the Property and survive any Transfer of the Property.

(g)   The Power Provider shall be solely responsible for the operation and maintenance of the Generating Facility, including without limitation the obligation to promptly make or pay (as determined by the Power Purchaser) for any repairs to the Property to the extent directly caused by the Power Provider, its employees, agents, contractors or subcontractors, and shall, at all times during the Term, maintain the Generating Facility in good operating condition. The Power Provider shall bear all risk of loss with respect to the Generating Facility, except for actions or negligence by the Power Purchaser or its agents and employees, and shall have full responsibility for its operation and maintenance in compliance with all laws, regulations and governmental permits. However, if such loss results in the cessation or reduction of the energy output by the Generating Facility, the Power Provider shall be relieved of its energy output guarantee obligations during such cessation or reduction of the energy output. The energy output guarantee shall recommence upon assumption of normal operation of the Generating Facility. The Power Provider shall coordinate in advance all such repair and maintenance work with the Site President or his/her designee so as not to restrict parking access or interfere with scheduled activities on the Site. Upon a request by the Power Provider that it conduct repair and maintenance work, the Power Purchaser shall respond to such request within ten (10) business days; if the Power Purchaser does not respond to such request within such ten (10) day period, such request shall be deemed approved by the Power Purchaser. All such work shall be diligently prosecuted to completion to the end that such work shall not remain in a partly finished condition any longer than necessary for completion.

(h)   In Operation, the Generating Facility is targeted to have a combined generating capacity rating as shown in Exhibit B.

(i)   In addition, if the Generating Facility must be moved to or placed at an alternate location at the Site during the Term, the alternate location is subject to the approval (such approval not to be unreasonably withheld or delayed) of the Power Purchaser, and, upon such approval, the obligations of the Parties remain as set forth in this Agreement. Notwithstanding the Generating Facility's presence as a fixture on the Site, the Party requiring such movement or replacement shall be responsible for all associated costs of removal and reinstallation. If the Power Purchaser requires movement or replacement, the Power Purchaser shall pay to the Power Provider, in addition to other amounts set forth in this Section 4(i), a monthly payment (prorated as needed) equal to the average power purchase set forth in Section 6 for the preceding twelve (12) months, or however long the Generating Facility has been in Operation if less than twelve (12) months, for the period of time during which the Generating Facility is not in Operation due to the movement or replacement.

(j) In addition, if temporary removal of the Generating Facility is required due to Site work unrelated to the Generating Facility, the Power Purchaser is responsible for all associated costs of removal and reinstallation and must proceed diligently. During any period while the Generating Facility is off-line in connection with a relocation, the Power Purchaser shall also pay the Power Provider a monthly payment (prorated as needed) equal to the average power purchase set forth in Section 6 for the preceding *[agreed period of time]*, or however long the Generating Facility has been in Operation if less than *[agreed period of time]*.

(k) Notwithstanding the Generating Facility's presence as a fixture on the Site, the Power Purchaser shall not cause or permit any interference with the Generating Facility's isolation and access to sunlight, as such access exists as of the Effective Date.

(l) The Power Purchaser will have the option to add-on additional electric power generation equipment to the Generating Facility; however, any such addition shall be separate and apart from this Project and require a separate agreement between the Parties, unless otherwise agreed to by each Party hereof. The Power Provider shall not be responsible or liable hereunder for any damage resulting from such addition by the Power Purchaser.

5. **POWER PROVIDER'S UNDERTAKING**

(a) The Power Provider shall immediately notify the Power Purchaser in writing when:

(i) the Power Provider merges with, acquires, or transfers all or substantially all its assets to another entity;

(ii) any person or entity acquires directly or indirectly the beneficial ownership of equity securities and, consequently, the power to elect a majority of the board of directors of the Power Provider, or otherwise acquires directly or indirectly the power to control the policy making decisions of the Power Provider;

(iii) the Power Provider is dissolved; applies for insolvency or bankruptcy; or otherwise admits in writing its inability to pay its outstanding obligations;

(iv) the Power Provider is administratively or judicially declared insolvent or bankrupt, placed under receivership, administration, rehabilitation, or liquidation;

(v)   the Power Provider's financial condition becomes significantly unstable and threatens to jeopardize the Power Provider's ability to perform its obligations under this Agreement;

(vi)   the Power Provider loses any license or authorization required to perform its obligations under this Agreement; or

(vii)   the Power Provider faces any event beyond the control of the Power Provider or a situation that makes it impossible for the Power Provider to carry out its obligations under this Agreement.

The Power Purchaser and the Power Provider shall explore alternative arrangements to implement this Agreement under any or all of the above circumstances.

(b)   Machines or equipment, if any, that the Power Purchaser issues to the Power Provider for free, shall remain the property of the Power Purchaser; and the Power Purchaser may recover them from the Power Provider at any time. The Power Provider shall not, under any circumstances, have a lien or any other interest on such machines or equipment; and the Power Provider shall at all times possess them only as fiduciary agent and bailee of the Power Purchaser. The Power Provider shall not commingle the machines or equipment with its own, and shall accordingly advise all sub-contractors and other interested third parties of the Power Purchaser's ownership of such machines or equipment.

(c)   The Power Provider shall compensate the Power Purchaser for the loss of or damage to machines or equipment that the Power Purchaser has provided to the Power Provider when the Power Purchaser finds that the loss or damage resulted from the willful act or gross negligence of the Power Provider's personnel. Upon expiration or termination of this Agreement, the Power Provider shall immediately return, without need of demand, the machine(s)/equipment that the Power Purchaser had supplied.

(d)   The Power Purchaser reserves the right to refuse entry into or remove from the Power Purchaser's Premises the Power Provider's personnel who, in the Power Purchaser's judgment, are under the influence of alcohol or other drugs or, for any reason, are deemed incapable of safely and reliably performing assigned work or whose behavior does not conform to generally accepted standards.

(e)   The Power Provider's personnel who commit an offense on the Power Purchaser's Premises shall be removed from the Power Purchaser's Premises and/or surrendered to local law enforcement authorities. For this purpose, the Power Provider recognizes the authority of [*name of department or person with authority*] to summon the Power Provider's employees for investigation. Such offenses include, but are not limited to, the following:

(i)   <u>Theft/Pilferage</u>   removing or attempting to remove from the Power Purchaser's Premises, without Gate Pass or authority to do so, the Power Purchaser's properties, regardless of the condition or value of such property; or stealing personal properties while on the Power Purchaser's premises.

(ii)   <u>Damage or Disruption</u>   deliberately or through culpable negligence disrupting the Power Purchaser's operations, and/or otherwise causing damage to or destroying the Power Purchaser's property.

(iii)   <u>Drunkenness/Alcoholism</u>   consuming intoxicating beverages on the Power Purchaser's Premises or reporting for work under the influence of alcohol.

(iv)   <u>Using Prohibited Drugs</u>   includes possessing, pushing, consuming or otherwise using prohibited drugs, hallucinogenic substances or narcotics on the Power Purchaser's premises.

(v)   <u>Gambling</u>   gambling in any form while on the Power Purchaser's premises.

(vi)   <u>Violence</u>   using force, physical assault, coercion, threat, intimidation, extortion, bribery, or engaging in other unlawful activities with Power Purchaser or non-Power Purchaser personnel for any purpose whatsoever.

(vii)   <u>Possessing Firearms and other deadly weapons</u>   carrying firearms, licensed or unlicensed, and/or other deadly weapons while on the Power Purchaser's premises.

(viii)   The Power Provider shall not pay any commissions, or fees; grant any rebates or give gifts or favors; or otherwise enter into any financial or business arrangements with the Power Purchaser's personnel or their dependents during the effectivity of this Agreement.

(f)   The Power Provider is an independent contractor of the Power Purchaser. This Agreement shall not nor be deemed to create the relationship of employer and employee, master and servant, or principal and agent between the Power Purchaser and the Power Provider or the Power Provider's employees, agents or any other persons engaged by the Power Provider to perform its obligations under this Agreement. Accordingly, neither Party shall be authorized to act in the name or on behalf of, or otherwise bind the other Party, save as expressly permitted by the terms of this Agreement.

6. PURCHASE AND SALE OF POWER - TAXES - TERM

6.1 Purchase and Sale

Beginning with the Operation of the Generating Facility and continuing throughout the Term, the Power Purchaser shall purchase and accept delivery from the Power Provider at the purchase price set forth in Exhibit C and the Power Provider shall sell and deliver to the Power Purchaser, the Energy Output (in such amount of output as the Generating Facility produces from time to time), but in no event, other than a Force Majeure event (as defined in Section 20), shall the Energy Output of the Generating Facility be less than the "Guaranteed Energy Output" described in part (ii) below. The Power Purchaser shall not resell any of the Energy Output.

(i) <u>Purchase Price</u>. The Power Purchaser shall pay to the Power Provider an amount equal to the Energy Output multiplied by the Purchase Price (per Exhibit C) per kWh AC. Such amount shall be paid in accordance with the terms of Section 10.

(ii) <u>Guarantee</u>. The Power Provider shall provide a Cumulative Output Guarantee from the Generating Facility (as set forth in Exhibit E) commencing on the date of Operation and continuing until the fifteenth (15th) anniversary of the Operation. The guarantee is defined to be **90%** of the expected annual production from the Generating Facility to be measured in kWh. In order to control for variations in weather, the actual output will be compared to the Cumulative Output Guarantee on a cumulative basis on the third (3rd), sixth (6th) ninth (9th), twelfth (12th) and fifteenth (15th) year during the cumulative output Guarantee Term. Actual production shall accrue to the cumulative balance each year and be compared on the anniversary dates noted above of the Operation Date to the aggregate cumulative output guarantee for the years in that measurement period as indicated in the table below. In the event that the Guaranteed Energy Output is not achieved as described above during the Term of this Agreement (the "Guaranteed Energy Output Shortage"), the Power Purchaser will have the right to purchase the generating facility, as per Section 6.3 (c) (ii) below. In such case, the purchase price shall be reduced to half of the then Fair Market Value or Buyout Value set forth in Exhibit D, whichever the greater.

(iii) <u>Certification</u>. The Power Provider shall provide a Plant Performance Certification issued by an independent third party international accreditation body, which shall be the basis of calculation of the purchased output.

(iv) <u>Purchase Price Revision in Exceptional Circumstances.</u> The Purchase Price may be revised under the terms of this Section to reflect exceptional variations in the costs of labor.

The price revision shall take place only once every three (3) years after the Operation Date, and shall be computed by reference to the minimum wage, as established by [*agreed reference index, e.g. the Government of the Philippines*] on the first day of Operation initially, and thereafter on each third anniversary date ("base minimum wage"). The revision will only take place if the minimum wage in force for the month preceding the relevant third anniversary ("current minimum wage") exceeds by 30% the previous base minimum wage.

For the purpose of the price revision, the proportion of the kWh Purchase Price attributable to labor costs shall be _____ percent, and the proportion attributable to non-labor costs of _____ percent.

The revised Purchase Price shall be determined by adding to the portion of the price attributable to non-labor costs the adjusted labor cost. The adjusted labor cost shall be obtained by multiplying _____ percent of the Purchase Price per kWh by a fraction, of which the numerator shall be the current minimum wage and the denominator shall be the base minimum wage. The current minimum wage used for the purpose of revising the Purchase Price shall become the base minimum wage to be taken into account for the next revision.

In any event, the total increase in the Purchase Price per kWh shall not exceed ten percent (10%) of the original Purchase Price per kWh, as defined in Exhibit C. The revised Purchase Price shall apply to the power supplied from the day following the relevant third anniversary.

6.2 Taxes

(a) The Power Provider shall be solely responsible for payment of taxes on its income.

The Power Provider and the Power Purchaser shall explore alternative arrangements to implement this Agreement, if any tax or duty other than taxes on net income is levied or if there is an attempt to levy any such duty or tax in connection with the performance of this Agreement.

(b) The Power Provider shall pay all real estate or personal property taxes, possessory interest taxes, business or license taxes or fees, service payments in lieu of such taxes or fees, annual or periodic license or use fees, excises, assessments, bonds, levies, fees or charges of any kind which are assessed, levied, charged, confirmed, or imposed by any public authority due to the Power Provider occupancy and use of the Premises or any portion or component thereof.

6.3 Term

(a)   The Term of this Agreement shall commence as of the Effective Date and shall expire at 2400 hours on the date fifteen (15) years following the Operation Date.

(b)   The Power Purchaser may not terminate this Agreement or require a substitute Site within the first six (6) years from the Operation Date, except as the result of an Event of Default (defined in Section 15) by the Power Provider. In the event the Power Purchaser determines in the Power Purchaser's sole discretion that the Premises are needed at any time on or after the expiration of the sixth (6th) year following the Operation Date, the Parties agree that the Generating Facility will be relocated, at the Power Purchaser's sole cost, expense and risk, and at no cost to the Power Provider, to a mutually agreed upon location on the Site. The Power Purchaser shall provide the Power Provider with not less than six (6) months prior written notice of the Power Purchaser's proposed relocation of the Generating Facility.

(c)   Upon expiration of this Agreement or termination by either Party pursuant to this section, the Power Purchaser may choose one of the following options:

(i)   Removal of Generating Facility. The Power Provider shall remove the Generating Facility from Site by a mutually convenient date but in no case later than one hundred eighty (180) days after such expiration or termination, subject to the Power Purchaser's reimbursement of the Power Provider's reasonable costs of removal (not to exceed *[agreed amount]*) if removal occurs during the Term of this Agreement as the result of the default of the Power Purchaser, or at the Power Provider's expense at the end of the Term or expiration of the Term. The Power Purchaser shall provide the Power Provider with reasonable access to perform such activities.

(ii)   Purchase of Generating Facility. Provided no default of the Power Purchaser shall have occurred and be continuing, the Power Purchaser may purchase the Generating Facility no earlier than seventy-two (72) months from the Operation Date. If the Power Purchaser elects to purchase the Generating Facility prior to the expiration of the Term, the purchase price shall be the greater of the then Fair Market Value or Buyout Value set forth in Exhibit D. If the Power Purchaser exercises the purchase option at the expiration of the Term, the Power Provider shall, consistent with the applicable laws of the *[territory]* take whatever actions are necessary to transfer fee title ownership to the Power Purchaser of the Generating Facility and all improvements placed by the Power Provider on the Premises, at no charge to the Power Purchaser, and free and clear from any lien or monetary encumbrance. Not less than ninety (90) days prior to the exercise of the purchase option, the Power Purchaser shall provide written notice to the Power Provider of the Power Purchaser's exercise thereof. Upon the exercise of the foregoing purchase option plus receipt of the then Fair Market Value or Buyout Value, as applicable, and all other amounts then owing by the Power Purchaser to the Power Provider, the Parties will execute all documents necessary to cause title to the Generating Facility to pass to the Power Purchaser as-is, where-is; provided, however, that the Power Provider shall remove any encumbrances placed on the Generating Facility by the Power Provider. The "Fair Market Value" of the Generating Facility shall be the value determined by the mutual agreement of the Power Purchaser and the Power Provider within ten (10) days of the Power Purchaser's termination notice pursuant to this Section 6. If the Power Purchaser and the Power Provider cannot mutually agree to a Fair Market Value, then the Parties shall select a nationally recognized independent appraiser with experience and expertise in the solar photovoltaic industry to value such equipment. Such appraiser shall act reasonably and in good faith to determine the Fair Market Value and shall set forth such determination in a written opinion delivered to the Parties. The valuation made by the appraiser shall be binding on the Parties in the absence of fraud or manifest error. The costs of the appraisal shall be borne by the Parties equally. To the extent transferable, the remaining period, if any, on all warranties for the Generating Facility will be transferred from the Power Provider to the Power Purchaser at the Power Purchaser's sole expense.

(d)   Termination Value. In the event that a termination occurs for reasons attributable to the Power Purchaser (including termination pursuant to Section 6.3(c), but not including a Force Majeure event which continues for at least one (1) year or a the Power Purchaser default pursuant to Section 15), the Power Purchaser shall pay to the Power Provider the greater of Fair Market Value or the Termination Value as set forth in Exhibit D (which shall be prorated for partial years), plus all other amounts then owing by the Power Purchaser to the Power Provider. The Parties further agree the Termination Value is not an approximation of the Fair Market Value. The Termination Value shall include the Power Provider's cost of removal.

(e)   Termination for Convenience. If the Power Provider is unable to assign this Agreement pursuant to the terms set forth in Section 25 below by or before [agreed date], the Power Provider shall have the right to terminate this Agreement for convenience upon written notice to the Power Purchaser. If the Agreement is so terminated, the Power Provider shall reimburse the Power Purchaser for legal costs it has incurred to date, in an amount not to exceed [agreed amount].

7.   **ENVIRONMENTAL ATTRIBUTES**

(a)   <u>Delegation of Attributes to the Power Provider</u>   Notwithstanding the Generating Facility's presence as a fixture on the Site, the Power Provider shall own, and may assign or sell in its sole discretion, all right, title and interest associated with or resulting from the development and installation of the Generating Facility or the production, sale, purchase or use of the Energy Output including, without limitation:

(i)   all Environmental Incentives except for Solar Renewable Energy Credits associated with the Generating Facility; and

(ii)   the Reporting Rights and the exclusive rights to claim that: (A) the Energy Output was generated by the Generating Facility; (B) the Power Provider is responsible for the delivery of the Energy Output to the Energy Delivery Point; (C) the Power Provider is responsible for the reductions in emissions of pollution and greenhouse gases resulting from the generation of the Energy Output and the delivery thereof to the Energy Delivery Point; and (D) the Power Provider is entitled to all credits, certificates, registrations, etc., evidencing or representing any of the foregoing.

(b)   <u>Delegation of Attributes to the Power Purchaser</u>   The Power Purchaser shall own, and may assign or sell in its sole discretion, all right, title and interest associated with or resulting from the following:

(i)   All Environmental Attributes and Solar Renewable Energy Credits associated with the Generating Facility; and

(ii)   the Reporting Rights and the exclusive rights to claim that the Power Purchaser is entitled to all Solar Renewable Energy Credits evidencing or representing any of the foregoing.

8.   **METERING**

(a)   The Power Provider shall install and maintain a standard revenue quality meter with electronic data acquisition system ("DAS") capabilities at the Generating Facility. The meter shall measure the alternating current output of the Generating Facility on a continuous basis. The Power Provider shall be responsible for maintaining the metering equipment in good working order and, if the Power Purchaser so requests, for testing at the Power Purchaser's sole expense the same once per calendar year and certifying the results of such testing to the Power Purchaser. In the event of a failure of the electronic meter reading system and until such failure has been corrected, the Power Provider shall be responsible for conducting monthly on-site readings of the standard electricity meter to determine the output of the Generating Facility delivered to the Power Purchaser. Data retrieved from any such meter shall serve as the basis for invoicing the Power Purchaser for all delivered energy.

(b)   The Power Provider shall maintain all DAS data and shall provide to the Power Purchaser a report of the Site's individual metered energy, as read and collected on a monthly basis, once each month within fourteen (14) business days after the last day of the preceding month. The Power Provider shall verify and adjust all DAS data at least once per calendar year based on readings from the foregoing standard meter. Subject to Section 8(a), such data, as verified and adjusted, shall serve as the basis for invoicing the Power Purchaser for all delivered energy. The Power Provider shall preserve all data compiled hereunder for a period of at least two (2) years following the compilation of such data.

(c)   As may be periodically requested by the Power Purchaser, the Power Purchaser shall have the right to audit all such DAS data upon reasonable notice, and any such audit shall be at the Power Purchaser's sole cost (unless an audit reveals at least a ten percent (10%) overcharge to the Power Purchaser, in which case the Power Provider shall bear the cost of that audit). The Power Purchaser shall have a right of access to all meters at reasonable times and with reasonable prior notice for the purpose of verifying readings and calibrations. If the metering equipment is found to be inaccurate, it shall be corrected and past readings shall be promptly adjusted in an equitable manner.

(d)   The Power Provider shall provide connection points for the Power Purchaser's Data Acquisition/Building/Energy Management Systems upon the Power Purchaser's request and at the Power Purchaser's sole cost and expense.

9.   **DELIVERY**

(a)   Title and risk of loss of the Energy Output shall pass from the Power Provider to the Power Purchaser upon delivery of the Energy Output at the Energy Delivery Point. All deliveries of Energy Output hereunder shall be in the form of three-phase, sixty-cycle alternating current or similar to properly integrate with the Site's electrical system. The Power Purchaser shall purchase and accept delivery of metered Energy Output at the Energy Delivery Point.

(b)   The Power Provider shall ensure that all energy generated by the Generating Facility conforms to Utility specifications for energy being generated and delivered to the Site's electric distribution system, which shall include the installation of proper power conditioning and safety equipment, submittal of necessary specifications, coordination of Utility testing and verification, and all related costs.

(c)   The Power Purchaser shall be responsible for arranging delivery of Energy Output from the Energy Delivery Point to the Power Purchaser and any installation and operation of equipment on the Power Purchaser's side of Energy Delivery Point necessary for acceptance and use of the Energy Output. The Parties acknowledge that adjustments in the terms and conditions of this

Agreement may be appropriate to account for rule changes in the respective Utility or Utility control areas, by the respective independent system operators, or their successors, that could not be anticipated at the date of execution of this Agreement or that are beyond the control of the Parties, and the Parties agree to make such commercially reasonable amendments as are reasonably required to comply therewith.

10. INVOICES AND PAYMENT

The Power Purchaser shall pay the Price, according to the agreed terms and manner of payment therein and subject to the following conditions:

(a)   The Power Provider's request(s) for payment shall be made to the Power Purchaser in writing; accompanied by an invoice by the [*insert day*] business day of each calendar month (or upon a monthly schedule reasonably acceptable to the Power Purchaser and the Power Provider), stating the Energy Output delivered to the Power Purchaser during the preceding calendar month and calculating the purchase price.

(b)   Payments shall be made only after the Head of the User Unit certifies that the Power delivered is found to be delivered according to the terms of this Agreement.

(c)   The Power Purchaser shall pay promptly and not later than thirty (30) days after the Power Provider submits an invoice. The Power Provider shall accept payments as full satisfaction of the Power Provider's entire claim arising out of or in connection with this Agreement.

(d)   Except with the prior approval of the Head of the User Unit, no payment shall be made for the power not yet delivered under this Agreement.

(e)   Invoices and payments schedule shall commence following Operation. The Power Provider in its sole discretion may bill the Power Purchaser the annual fixed capacity charge in either prorated monthly amounts or as an annual fee.

(f)   Unless otherwise agreed with the Power Provider, the Power Purchaser shall remit payment of the Price by electronic transfer. The Power Provider shall provide the Power Purchaser its bank details such as bank name, bank address/branch, account name, and account number on its invoices or request for payment.

11. INVOICE ADJUSTMENTS - DISPUTES OVER INVOICES

Either Party may, in good faith, dispute the correctness of any invoice or any adjustment to an invoice rendered or adjust any invoice for any arithmetic, computational or meter-related error within twelve (12) months of the date the invoice or adjustment to an invoice was rendered. In the event a Party disputes all or a portion of an invoice, or any other claim or adjustment arises, that Party shall pay the undisputed portion when due and provide the other Party notice of the dispute and the amount in dispute. In such event, the Parties shall first use good faith, reasonable, diligent efforts to resolve such dispute within a reasonable period of time not to exceed thirty (30) days from the date of such notice. If the Parties do not resolve such a dispute within such thirty (30) days, then the Parties may pursue their rights appropriately. The Power Purchaser shall pay to the Power Provider any disputed amount which is ultimately determined to have been properly billed to the Power Purchaser, together with interest at a rate of 1% per month or as set forth in the *[appropriate law or policy]*, (whichever amount is greater) on any disputed amount which is ultimately determined to have been properly billed to the Power Purchaser, until such properly billed amount is paid, which shall be the Power Provider's sole and exclusive remedy with respect to a dispute concerning any invoice.

12. WARRANTIES AND REPRESENTATIONS

(a) The Power Provider warrants that it has full capacity, authority and consent, including the consent of its parent company, where applicable, and that it possesses the necessary licenses, permits, and power to execute and perform its obligations under this Agreement. The Power Provider further warrants that this Agreement is executed by the Authorized Representative of the Power Provider.

(b) All information contained in the Power Provider's Bid is true, accurate and not misleading, except those that the Power Provider may have specifically disclosed in writing to the Power Purchaser before executing this Agreement;

(c) To the best of the Power Provider's knowledge and belief, no claim is being asserted and no litigation, arbitration or administrative proceeding is presently in progress, pending or being threatened against the Power Provider or any of its assets that could materially and adversely affect the Power Provider's ability to deliver the power and related services under this Agreement.

(d) The Power Provider is not subject to any contractual obligation that would materially and adversely affect the Power Provider's ability to deliver the power and related services under this Agreement; nor has the Power Provider done or omitted to do anything that could materially and adversely affect its assets, financial condition or position as a going business concern.

(e) The Power Provider has not filed nor is it facing proceedings for winding up its business or for dissolution, insolvency, bankruptcy, or the appointment of a receiver, liquidator, administrator or similar officer in relation to any of the Power Provider's assets or revenue. The Power Provider expressly warrants its financial viability and shall permit the Power Purchaser to inspect the Power Provider's accounts, financial statements and other records relevant to the performance of the Power Provider under this Agreement, or otherwise have these accounts and records audited externally, as the Power Purchaser may deem necessary.

(f)   The Power Provider has undertaken all financial accounting and reporting activities required under the generally accepted accounting principles that apply to the Power Provider and in the country where it is registered and has complied with applicable securities and tax laws and regulations.

(g)   The Power Purchaser represents and warrants to the Power Provider that there are no circumstances known to the Power Purchaser and commitments to third parties that may damage, impair, or otherwise adversely affect the Generating Facility or its function (including activities that may adversely affect the Generating Facility's exposure to sunlight).The Power Purchaser covenants that the Power Purchaser has lawful title to the Property and the Premises and full right to enter into this Agreement and that, subject to the Power Provider's compliance with all material provisions contained in this Agreement, the Power Provider shall have quiet and peaceful possession of the Premises throughout the term of this Agreement. The Power Purchaser will not initiate or conduct activities that it knows or reasonably should know may damage, impair, or otherwise adversely affect the Generating Facility or its function (including activities that may adversely affect the Generating Facility's exposure to sunlight), without the Power Provider's prior written consent, which consent shall not be unreasonably withheld or delayed.

13.   COVENANTS

(a)   <u>Security</u>. The Power Purchaser shall provide and take reasonable measures for security of the Generating Facility.

(b)   <u>Liens</u>. Notwithstanding the Generating Facility's presence as a fixture on the Site,the Power Purchaser shall not directly or indirectly cause, create, incur, assume or suffer to exist any mortgage, pledge, lien (including mechanics', labor, or material man's lien), charge, security interest, encumbrance, or claim on or with respect to the Generating Facility or any interest therein. If the Power Purchaser breaches its obligations under this Section 13(b), it shall immediately notify the Power Provider in writing, shall promptly cause such liens to be discharged and released of record without cost to the Power Provider, and shall indemnify the Power Provider against all costs and expenses (including reasonable attorneys' fees and court costs at trial and on appeal) incurred in discharging and releasing such liens.

14.   PERFORMANCE SECURITY - GSMC – INDEMNIFICATION - INSURANCE

14.1   Performance Security

(a)   To secure performance of its obligations under this Agreement, the Power Provider shall post sufficient security in the amount of USD [*security amount*].

The proceeds of the Performance Security shall be payable to the Power Purchaser as compensation for any loss resulting from the Power Provider's failure to complete its obligations under this Agreement.

(b) The Performance Security shall be denominated in US Dollars or in a freely convertible currency acceptable to the Power Purchaser. The Performance Security shall be in one of the following forms:

    (i) Cash, cashier's check, manager's check, or bank draft;

    (ii) Bank guarantee or an irrevocable or stand-by letter of credit issued by a reputable bank in [*country*] or abroad and acceptable to the Power Purchaser;

The Performance Security is USD [*security amount*].
The Performance Security shall be in the form of _____.
The Performance Security shall be submitted on _____.
The Power Purchaser shall discharge the Performance Security on _____.

(c) The Power Provider shall submit the performance security within fifteen (15) calendar days from receipt of the Notice of Award from the Power Purchaser and in no case later than the signing of this Agreement by both parties.

(d) The Power Purchaser shall discharge and return the performance security to the Power Provider not later than thirty (30) days following the Operation Date.

(e) The Power Purchaser shall have the right to unilaterally call the Performance Security when the Power Purchaser determines that:

    (i) The Power Provider, in violation of or contrary to its warranties under this Agreement, does not have the required license, permit, power and/or authority to enter into and fully perform its obligations under this Agreement; or

    (ii) The Power Provider breached this Agreement and failed to remedy the breach, if the Power Purchaser deems such breach remediable, within seven (7) calendar days from receipt of notice from the Power Purchaser; or

    (iii) The Power Provider failed to submit to the Power Purchaser the Good Social Management Certificate included among the Contract Documents; misrepresented or made false statements in the Certificate; or failed to fulfill any of its warranties and obligations under this Agreement.

14.2 Good Social Management Certificate (GSMC)

(a) The Power Provider shall perform its obligations under this Agreement diligently, observe good social management practices, and comply with relevant laws, regulations, decrees and orders of competent government agencies or authorities concerning the employees of the Power Provider.

(b) The Power Provider shall submit to the Power Purchaser the Good Social Management Certificate (GSMC) within fifteen (15) days from the effective date of Agreement and subsequently on an basis during the effectivity of this Agreement.

(c) The Power Provider shall indemnify and hold the Power Purchaser free and harmless from any and all claims made by the Power Provider's personnel under [territory] labor laws and other related legislation, including but not limited to, the minimum wage law.

14.3 General Conditions for Performance Security, Insurance, and GSMC

(a) The Power Provider shall furnish evidence that the securities, insurance and/or GSMC were taken at the time required and continues to be in effect before executing this Agreement; during the Term and for one year after termination. The Power Provider shall deposit with the Power Purchaser a copy of the required GSMC, securities, insurance policy/ies, and receipts for payment of the corresponding premium, as applicable, within the time required above.

(b) The Power Provider shall obtain and maintain such GMSC, insurance and/or performance securities as the Parties may agree.

(c) When the Power Provider fails to submit, validate and/or renew any or all of the following contract requirements, namely: (a) Performance Security, (b) Comprehensive General Liability Insurance and (c) Good Social Management Certificate, the Power Purchaser reserves the right to withhold payment attributed to the Price until the aforesaid requirement/s is/are provided.

(d) The Power Provider shall not pass on under whatever form the penalty referred to in Subparagraph 14.3 (c) above or any part thereof to its employees.

(e) [*Name of department or person*], in coordination with the User Unit, shall monitor compliance with these requirements.

14.4 Indemnification and insurance

(a) The Power Provider and the Power Purchaser (each, in such case, an "Indemnifying Party.") shall indemnify, defend and hold the other Party and its employees, directors, officers, managers, members, shareholders and agents (each, in such case, an "Indemnified Party") harmless from and against any and all third party claims, suits, damages, losses, liabilities, expenses, and costs (including reasonable attorney's fees) including, but not limited to, those arising out of property damage (including environmental claims) and personal injury and bodily injury (including death, sickness and disease) to the extent caused by the Indemnifying Party's (i) material breach of any obligation, representation or warranty contained herein and/or (ii) negligence or willful misconduct.

(b)   The Power Provider shall maintain during the Term of this Agreement, with the Power Purchaser named as additional insured therein as its interest may appear, for the duration of this Agreement the insurance coverage outlined in (i) through (iii) below, and all such other insurance as required by applicable law. Evidence of coverage will be provided to the Power Purchaser on an annual basis, prior to policy expiration, via a Certificate of Insurance or a Self Administered Claims Letter, Commercial General Liability insurance applying to the use and occupancy of the Premises, and the business operated by The Power Provider thereon in the following amounts:

(i)   Commercial General Liability Limits:

$[*liability limit*] General Aggregate

$[*liability limit*] Products & Completed Operations Aggregate

$[*liability limit*] Each occurrence

$[*liability limit*] Personal Injury (Advertising Injury excluded)

$[*liability limit*] Fire Damage, Any One Fire

$[*liability limit*] Medical Payments, Each Person

(ii)   Excess Liability

Limits: $[*liability limit*] Each occurrence

$[*liability limit*] Aggregate

(iii)   Standard for property insurance ("All Risk" coverage) equal to at least 90% of the replacement cost covering the Generating Facility, and all other improvements placed by the Power Provider on the Premises.

Any policy or policies of worker's compensation, extended coverage or similar casualty insurance, which either Party obtains in connection with the Premises, shall include a clause or endorsement denying the insurer any rights of subrogation against the other party to the extent the insured party has waived rights of recovery against such other party prior to the occurrence of injury or loss. The Power Provider and the Power Purchaser waive any rights of recovery against the other for injury or loss due to hazards covered by insurance obtained under this Agreement. Within thirty (30) days after execution of this Agreement and upon the Power Purchaser's request annually thereafter, the Power Provider shall deliver to the Power Purchaser true and correct copies of all certificates of insurance evidencing such coverage. These certificates shall specify that the Power Purchaser shall be given at least thirty (30) days prior written notice by the insurer in the event of any material modification, cancellation or termination of coverage. Such insurance shall be primary coverage without right of contribution from any insurance of the Power Purchaser. Should any such policy of insurance be cancelled or changed, the Power Provider agrees to immediately provide the Power Purchaser true and correct copies of all new or revised certificates of insurance.

(c)   Any insurance maintained by the Power Purchaser is for the exclusive benefit of the Power Purchaser and shall not in any manner inure to the benefit of the Power Provider, except to the extent that any payments made for claims related to the loss or damage of the Generating Facility owned by the Power Provider.

(d)   If the Generating Facility is (i) materially damaged or destroyed, or suffers any other material loss or (ii) condemned, confiscated or otherwise taken, in whole or in material part, or the use thereof is otherwise diminished so as to render impracticable or unreasonable the continued production of energy, to the extent there are sufficient insurance or condemnation proceeds available to the Power Provider, the Power Provider shall either cause (A) the Generating Facility to be rebuilt and placed in Operation at the earliest practical date or (B) another materially identical Generating Facility to be built in the proximate area of the Site and placed in Operation as soon as commercially practicable.

(e)   All property insurance shall be the responsibility of the Power Purchaser. The Power Purchaser represents that it maintains and covenants that it shall maintain during the term of this Agreement property insurance sufficient to insure it against complete loss or destruction of the Property, including losses occasioned by operation of the Generating Facility, whether or not involving the fault of the Power Provider.

15.   TERMINATION FOR DEFAULT

The Power Purchaser shall terminate this Agreement for default when:

(a)   The Power Provider fails to or incurs delay in supplying the guaranteed power amount;

(b)   The undelivered power amounts to at least ten percent (10%) of the annual purchase price;

(c)   The Power Provider fails to continue delivering power amounting to at least ten percent (10%) of the annual purchase price within sixty (60) calendar days from receipt of written notice from the Power Purchaser informing the Power Provider that the Force Majeure had ceased; or

(d)   The Power Provider fails to perform any other obligation under this Agreement.

(e)   When the Power Purchaser terminates this Agreement in whole or in part, the Power Purchaser may procure power from solar origin and similar services, and the Power Provider shall be liable for any excess costs that the Power Purchaser may incur as a result. The Power Provider shall continue performing its obligations provided under parts of this Agreement that remain effective.

16.   TERMINATION FOR INSOLVENCY AND CHANGE OF CONTROL

(a)   The Power Purchaser may terminate this Agreement when:

(i)   the Power Provider undertakes legal proceedings to dissolve or wind up its business, or be declared bankrupt and/or insolvent; or

(ii)   a creditor or encumbrancer attaches or takes possession of, or a distress, execution, sequestration or other such process is levied or enforced on or sued against, the whole or any part of the Power Provider's assets and such attachment or process is not discharged within fifteen (15) days.

(b)   Termination for insolvency shall not entitle the Power Provider to compensation other than for the power already delivered; it shall be without prejudice to any right of action or remedy that has accrued or will accrue thereafter to the Power Purchaser and/or the Power Provider.

(c)   If a significant change in the ownership and/or control of the Power Provider threatens to disrupt or adversely affect delivery of power, the Power Purchaser may terminate this Agreement for change of control when:

(i)   the Power Provider merges with, acquires, or transfers all or substantially all its assets to another entity;

(ii)   any person or entity acquires directly or indirectly the beneficial ownership of the Power Provider and, consequently, the power to elect a majority of the board of directors of the Power Provider;

(iii)   any person or entity otherwise acquires directly or indirectly the power to control the policy making decisions of the Power Provider; or

(iv)   where applicable, the Power Provider dies or otherwise loses legal capacity to contract.

17.   TERMINATION FOR UNLAWFUL ACTS

The Power Purchaser may terminate this Agreement if the Power Purchaser determines that the Power Provider has committed unlawful acts during the provision of solar power and Related Services or implementation of this Agreement. Unlawful acts include, but are not limited to, the following:

(a)   corrupt, fraudulent, or coercive practices defined in Section 34 below;

(b)   forging or using forged documents;

(c)   using adulterated materials, means or methods; or using production methods contrary to the rules of science or the trade;

(d)   any of the offenses enumerated in Section 5(e) above; or

(e)   other acts analogous to the foregoing.

18.   PROCEDURES FOR TERMINATION

(a)   [*Name of department or person*], on its own or, within seven (7) days upon receipt of a written report from the Head of the User Unit alleging acts or causes that may constitute ground(s) for termination, shall verify the existence of ground(s) for termination.

(b)   [*Name of department or person*] shall submit to [*name of department or person*] a Verified Report with supporting documents or evidence and a corresponding recommendation to commence termination. Upon approval of [*name of department or person*] shall send to the Power Provider a copy of the Verified Report and a written notice stating:

(i)   that this Agreement is being terminated for the ground(s) mentioned above, with summary statement of the acts/omissions that constitute the ground(s) for terminating;

(ii)   the extent of termination, whether in whole or in part;

(iii)   that the Power Provider must show cause why this Agreement should not be terminated; and

(iv)   special instructions of the Power Purchaser, if there are any.

(c)   Within seven (7) calendar days from receipt of the notice, the Power Provider shall submit to [*name of department or person*] an answer stating why this Agreement should not be terminated. If the Power Provider fails to answer, or [*name of department or person*], in consultation with the User Unit, deems the answer unacceptable, [*name of department or person*] shall recommend termination to [*name of department or person*].

(d)   After evaluating the Power Provider's answer and the Verified Report, [*name of department or person*] shall endorse termination, as appropriate, to the Power Purchaser's approving Authority. Within a non-extendable period of ten (10) calendar days from receipt of the endorsement, the approving authority of this Agreement shall decide to terminate this Agreement or not. The Power Purchaser shall serve a written notice to the Power Provider of the decision and, unless otherwise provided in the notice, this Agreement shall be deemed terminated immediately upon the Power Provider's receipt of the notice.

(e)   [*Name of department or person*] or [*name of department or person*], as appropriate, may create a Contract Termination Review Committee (CTRC) to assist in evaluating cases for termination. Decisions recommended by the CTRC shall be subject to the approval of the applicable approving authority.

19.   TERMINATION AND LIQUIDATED DAMAGES

(a)   In case the unexcused delay in completion of the installation work exceeds a time duration equivalent to twenty-five percent (25%) of the time specified in Section 4(b) plus any extension duly granted to the Power Provider, the Power Purchaser may forfeit the Power Provider's performance security and take over the prosecution of the project or award the same to a qualified Power Provider through negotiated contract.

(b)   After the Operation Date, upon an Event of Default by one Party, the other Party shall have the right, but not the obligation, to terminate or suspend this Agreement with respect to all obligations arising after the effective date of such termination or suspension (other than payment obligations relating to obligations arising prior to such termination or suspension). The Parties acknowledge that given the complexity of this technology and the volatility of energy markets, adequate damages in the event of breach of contract will be difficult if not impossible to calculate. Consequently, the Parties agree that in the event of a default under this Agreement that leads to termination, the non-defaulting Party may pursue all remedies available in accordance with this Agreement and the defaulting Party's liability hereunder shall be determined as follows:

(i)   The Power Provider's liability after Operation Date, shall be equal to the product of (A) the purchase price per kWh AC of Energy Output, as set forth under Section 6 pursuant to Exhibit C times (B) the number of days remaining in the term of the Agreement times the daily Average kWh AC Output (as defined below); or

(ii)   as to the Power Purchaser's liability, an amount equal to the costs of removing the Generating Facility, the present value of the Power Purchaser's purchase obligations hereunder with respect to the Energy Output of the Generating Facility for the remaining term of the Agreement, and during the term of this Agreement, the value of any Environmental Incentives and Environmental Attributes (including any applicable tax credits) that would have accrued to the Power Provider if the default did not occur. These damages shall be calculated as follows: number of days remaining in the Term of the Agreement then in effect times the product of (x) the purchase price per kWh AC the Power Purchaser would otherwise pay for such Energy Output pursuant to Section 6 (as such Purchase Price would have been escalated over time pursuant to Exhibit C times (y) the Average kWh AC Output. For purposes of calculating damages, "Average kWh AC Output" means the daily average number of kWh AC of energy actually delivered to the Power Purchaser from the applicable Generating Facility beginning on the start of Operation through the date of the Power Purchaser's default. If a the Power Purchaser's default occurs prior to the completion of the first one (1) full year after the start of Operation of the Generating Facility, for purposes hereof, it shall be assumed that the "Average kWh Output" of the Generating Facility during such partial year of Operation was the expected daily number of kWh AC of Energy Output, calculated by dividing the "Estimated Annual 1st Year Production" as set forth by the Generating Facility in Exhibit A by 365 days. The Power Purchaser's liability for such liquidated damages may be partially mitigated to the extent that the Power Provider is able to enter into alternative arrangements with another power purchaser to install the Generating Facility at another site and sell its energy output to the substitute power purchaser on equal or superior terms than stated in this Agreement.

(iii)   In either case, the defaulting Party shall be liable to reimburse the non-defaulting Party for such non-defaulting Party's expenses and costs relating to such default (including but not limited to reasonable attorney's fees).

20.   FORCE MAJEURE

(a)   In the event that either Party is delayed in or prevented from performing or carrying out its obligations under this Agreement by reason of any cause beyond the reasonable control of, and without the fault or negligence of, such Party (an event of "Force Majeure"), such circumstance shall not constitute an event of default, and such Party shall be excused from performance hereunder and shall not be liable to the other Party for or on account of any loss, damage, injury, or expense resulting from, or arising out of, such delay or prevention; provided, however, that the Party encountering such delay or prevention shall use commercially reasonable efforts to remove the causes thereof (with failure to use such efforts constituting an event of default hereunder). The settlement of strikes and labor disturbances shall be wholly within the control of the Party experiencing that difficulty.

(b)   As used herein, the term "Force Majeure" shall include, without limitation, (i) sabotage, riots or civil disturbances, (ii) acts of God, (iii) acts of the public enemy, (iv) terrorist acts affecting the Site, (v) an annual level of direct beam solar resource availability that is less than or equal to 90% of historical averages as measured by long-term weather data (minimum of five [5] years) collected at the Site and/or other reliable calibrated and appropriate weather station representative of the Site, (vi) volcanic eruptions, earthquake, hurricane, flood, ice storms, explosion, fire, lightning, landslide or similarly cataclysmic occurrence, (vii) or any action by any governmental authority which prevents or prohibits the Parties from carrying out their respective obligations under this Agreement. Economic hardship of either Party shall not constitute a Force Majeure under this Agreement.

(c)   In cases of Force Majeure, the Power Provider shall promptly notify the Power Purchaser in writing of the relevant circumstances. Unless otherwise directed by the Power Purchaser in writing, the Power Provider shall continue performing as much of its obligations as reasonably practical, and undertake reasonable alternative means of performance not prevented by the Force Majeure.

21.   RECORDS

Each Party hereto shall keep complete and accurate records of its operations hereunder and shall maintain such data as may be necessary to determine with reasonable accuracy any item relevant to this Agreement. Each Party shall have the right to examine all such records insofar as may be necessary for the purpose of ascertaining the reasonableness and accuracy of any statements of costs relating to transactions hereunder.

22.   NOTICES AND COMMUNICATION

(a)   Notices and communication required under this Agreement, including, but not limited to, requests, permissions or consent, shall be in writing. Notices and communication may be personally exchanged, sent in electronic format or by traditional means of communication such as registered mail, telex, telegram, or facsimile.

(b)   Notices shall be effective when sent to the address specified below and personally received by the addressee or constructively through the addressee's duly authorized representative. Notices sent by registered mail shall be effective on the date of delivery, as shown in the return card for registered mail or the postmaster's certification. Otherwise, notices sent by telex, facsimile or similar means shall be effective upon successful transmission to the Party or on the notice's effective date, whichever is later. All such communications shall be mailed, sent or delivered, addressed to the Party for whom it is intended, at its address set forth below:

(c)   The addresses are:

The Power Purchaser: _____

Attention: _____

Telephone: _____

Facsimile: _____

The Power Provider: _____

Attention: _____

Telephone: _____

Facsimile: _____

23.   AUTHORIZED REPRESENTATIVE

(a)   For purposes of giving Notices or communicating with each other, the contact details and authorized representative of the Parties are:

For the Power Purchaser: _____

_____

_____

_____

For the Power Provider: _____

_____

_____

Either Party may designate a new Authorized Representative by serving written notice on the other. The designation shall take effect immediately upon receipt of the Notice.

(b)   Orders, directives, and instructions given on behalf of the Power Purchaser to the Power Provider shall be communicated by the [*name of department or person*], or an officer duly designated under the Power Purchaser's rules and notified in advance to the Power Provider.

(c)   The Head of the User Unit is the designated officer in charge of monitoring the Power Provider's performance and shall recommend, among others, proper disposition of technical issues in implementing this Agreement. Communications regarding these technical issues shall be addressed to the Head of the User Unit in the Power Purchaser.

24.   CONFIDENTIALITY

All non-public information (including the terms of this Agreement and, in particular, the purchase price set forth in Section 6 and Exhibit C hereof) provided by either Party to the other or which is identified by the disclosing Party in writing as confidential or proprietary information shall be treated in a confidential manner and shall not be disclosed to any third party without the prior written consent of the non-disclosing Party, which consent shall not be unreasonably withheld. Notwithstanding the preceding, this Section and the restrictions herein contained shall not apply to any data or documentation which is:

(a)   required to be disclosed pursuant to a law, an order or requirements of a regulatory body or a court, after five (5) business days notice of such intended disclosure is given by the disclosing Party to the non-disclosing Party or if five (5) business days notice is not practical, then such shorter notice as is practical;

(b)   disclosed by a Party to an affiliate of such Party or in connection with an assignment permitted by Section 25; or

(c)   as of the time of disclosure, public knowledge without the fault of the disclosing Party.

25.   TRANSFER AND SUBCONT RACTING

(a)   The Power Provider shall not assign or transfer this Agreement or specific rights or obligations under it without the Power Purchaser's prior written consent.

(b)   When allowed by the Power Purchaser and by the type of works or services to be provided, and subject to the conditions under applicable legislations, the Power Provider may engage a Subcontractor or a Consultant to perform specific tasks under this Agreement. The relevant provisions of this Agreement shall apply to the Subcontractor, Consultant or their respective employees, as if they were employees of the Power Provider. However, the Power Provider shall be solely liable to the Power Purchaser for the provision the services delivered by the Subcontractor or Consultant.

(c)   The Power Purchaser may require the Power Provider to submit copies of such sub-contracting and/or consultancy agreements.

26.   SET-OFF

Except as otherwise set forth herein, each Party reserves to itself all rights, set-offs, counterclaims and other remedies and/or defenses to which it is or may be entitled, arising from or out of this Agreement or arising out of any other contractual arrangements between the Parties. All outstanding obligations to make, and rights to receive, payment under this Agreement may be offset against each other.

27.   BINDING EFFECT

The terms and provisions of this Agreement, and the respective rights and obligations hereunder of each Party, shall be binding upon, and inure to the benefit of, the Parties and their respective successors and permitted assigns.

28.   AMENDMENTS

(a)   The Power Purchaser and the Power Provider shall not vary or modify the terms of this Agreement except by prior written amendment signed duly executed by the parties.

(b)   The Power Provider shall submit [*name of department or person*], a written proposal to amend and/or modify this Agreement. Proposals to amend may include, but not be limited to, changes in the scope of the services, payment terms or completion schedule. The proposed amendment and/or modification shall not take effect until endorsed by [*name of department or person*] and approved by the appropriate approving authority.

29.   COUNTERPARTS

Any number of counterparts of this Agreement may be executed and each shall have the same force and effect as the original. Facsimile signatures shall have the same effect as original signatures and each Party consents to the admission in evidence of a facsimile or photocopy of this Agreement in any court or arbitration proceedings between the Parties.

30.   ENTIRETY AND SEPARABILITY

This Agreement supersedes all prior written or verbal agreement between the Power Purchaser and the Power Provider and contains the reciprocal obligations of the parties pertaining to or arising out of the delivery of solar power. However, this shall not excuse any Party from liability arising from fraud or fraudulent misrepresentation. Should any Paragraph, Subparagraph or part of this Agreement be held by a competent court or tribunal to be invalid, unenforceable, or void, the decision shall not affect the validity of the entire Contract or of those parts that are not so declared or otherwise remain capable of partial or separable performance.

31. GOVERNING LAW AND LANGUAGE

(a) This Agreement shall be governed and interpreted according to [*country*] law.

(b) English shall be the binding and controlling language on matters relating to the meaning and/or interpretation of this Agreement. Notices and other correspondences pertaining to this Agreement that the parties would exchange shall likewise be in English.

32. COOPERATION

Upon the receipt of a written request from the other Party, each Party shall execute such additional documents (e.g., Utility interconnection agreement), instruments and assurances and take such additional actions as are reasonably necessary and desirable to carry out the terms and intent hereof. Neither Party shall unreasonably withhold, condition or delay its compliance with any reasonable request made pursuant to this Section. Without limiting the foregoing, the Parties acknowledge that they are entering into a long-term arrangement in which the cooperation of all of them will be required.

33. SETTLEMENT OF DISPUTES

(a) The Power Purchaser and the Power Provider shall exert efforts to amicably resolve by mutual consultation disputes arising between them in connection with or as a result of this Agreement within thirty (30) days of either Party's notice of the dispute to the other. During this period, the User Unit, in consultation with [name of department or person], and the Power Provider's personnel directly involved should first attempt in good faith to settle the dispute among themselves before escalating it to [name of department or person] and their respective counterpart/s.

(b) After the initial thirty (30) day-period, the Parties shall consider referring unresolved disputes to mediation, unless the Power Purchaser considers the dispute not suitable for mediation or the Power Provider does not consent. The Parties shall appoint a neutral mediator from a reputable association of accredited mediators or their own short-list of dispute resolution professionals. The mediator shall formulate a simplified procedure for mediation and complete the mediation within fifteen (15) days from his appointment.

(c) Should efforts to resolve disputes under the preceding Subparagraphs fail, either party shall commence arbitration by sending notice to the other party stating in detail the issue to be resolved and that the dispute shall be referred to arbitration. The International Chamber of Commerce's (ICC) Rules of Arbitration in force upon commencement of arbitration shall apply. The arbitration shall be in English; it shall take place in Singapore and be governed by the laws of England and Wales. Each party shall pay its own costs.

(d)   Notwithstanding unresolved disputes, the Parties shall continue to perform their respective obligations under this Agreement or otherwise adopt provisional measures to ensure uninterrupted delivery of solar power.

34.   CORRUPT, FRAUDULENT, COERCIVE, AND COLLUSIVE PRACTICES[1]

(a)   For the purposes of this Section, the terms above shall have the following meaning:

    (i)   "Corrupt practice" is the offering, giving, receiving, or soliciting, directly or indirectly, anything of value to influence improperly the actions of another party.

    (ii)   "Fraudulent practice" is any act or omission, including a misrepresentation, that knowingly or recklessly misleads, or attempts to mislead, a party to obtain a financial or other benefit or to avoid an obligation.

    (iii)   "Coercive practice" is impairing or harming, or threatening to impair or harm, directly or indirectly, any party or the property of the party to influence improperly the actions of a party.

    (iv)   "Collusive practice" is an arrangement between two or more parties designed to achieve an improper purpose, including influencing improperly the actions of another party.

(b)   The Power Purchaser, bidders, manufacturers, or distributors, and the Power Provider shall observe the highest standard of ethics during the institutional procurement exercises of the Power Purchaser and implementation of this Agreement.

(c)   The Power Purchaser's Anticorruption Policy requires contractors under the Power Purchaser-financed contracts, as well as their staffs observe the highest ethical standards. Firms, entities and individuals bidding for or participating in the Power Purchaser's institutional procurement of Services and related goods, including but not limited to, service contractors and concessionaires, and their respective officers, employees and agents should report to [name of department or person] suspected acts of fraud or corruption that they come to know during the bidding process and throughout negotiation or execution of a contract.

(d)   Pursuant to its Anticorruption policy, the Power Purchaser:

    (i)   will not award a procurement contract to a winning bidder that has directly or in directly engaged in any corrupt, fraudulent, collusive, or coercive practice in competing for this Agreement in question;

---

[1] Power Purchasers may wish to replace this section with their own corruption policies.

(ii)   may suspend the procurement process at any stage when there is sufficient evidence to support a finding that an employee, agent or representative of the bidders, service contractors and concession-aires has engaged in any corrupt, fraudulent, collusive, or coercive practice in competing for, or in executing a Power Purchaser-financed contract;

(iii)   will sanction a bidder, service contractor, concessionaire or its suc-cessor, if the Power Purchaser at any time determines that such bid-der, service contractor, concessionaire or its successor has, directly or indirectly, engaged in any corrupt, fraudulent, collusive, or coer-cive practice in competing for, or in executing, any contract for the institutional procurement of services. Sanctions include, but are not limited to, declaring such bidder, service contractor, concessionaire or its successor ineligible to participate in Power Purchaser-financed activities indefinitely or for a stated period of time except under such conditions as the Power Purchaser deems appropriate; or reimburse-ment to the Power Purchaser of costs associated with investigations and proceedings.

(iv)   will take appropriate actions to manage conflicts of interest includ-ing, but not limited to, rejecting a proposal for award if it determines that a conflict of interest has flawed the integrity of any procurement process.

(e)   The Power Provider agrees to be bound by the Power Purchaser's Anticor-ruption Policy as outlined above.

(f)   The Power Provider shall permit the Power Purchaser to inspect the Power Provider's accounts and records relating to the performance of the Power Provider and to have them audited by auditors appointed by the Power Pur-chaser, if so required by the Power Purchaser.

35.   CONFLICT OF INTEREST

(a)   The Power Provider shall take appropriate steps to ensure that neither the Power Provider nor its personnel is placed in a position where, in the reason-able opinion of the Power Purchaser, there is or may be an actual or potential conflict between the pecuniary or personal interests of the Power Provider and performance of the Power Provider's obligations under this Agreement. The Power Provider shall disclose to the Power Purchaser full particulars of any such conflict of interest which may arise.

The Power Purchaser shall undertake measures to manage actual or potential conflicts of interest, including termination of this Agreement, as circumstances may warrant. This is without prejudice to other remedies or rights of action which shall have accrued or shall thereafter accrue to the Power Purchaser under this Agreement.

During and twelve (12) months after the Agreement Period, the Parties shall not employ or offer employment to any of the other Party's personnel who have been associated with the procurement and/or management of this Agreement without that other Party's prior written consent.

IN WITNESS WHEREOF, the undersigned have duly executed and delivered this Solar Services and Site Lease Agreement as of the day and year first above written.

The Power Provider

_____

By:

Name:

Title:

The Power Purchaser

_____

By:

Name:

Title:

APPENDICES

Exhibit A

Site Plan and Legal Description of Premises

The Power Provider will install a [*basic equipment specifications and type*] at the Site. The system will consist of: [*provide equipment or installation description*] constructed on the Premises, plus associated electrical equipment and monitoring devices.

Exhibit B

Estimated Annual Production for Agreement Term

Year	Expected Solar Electricity Produced (kWh)
1	
2	
3	
4	
5	
6	
7	
8	
9	
10	
11	
12	
13	
14	
15	
Total	

Note:
Guaranteed output level is at 90% of expected production level, for Year 1 to Year 15.

Exhibit C

Purchase Price

Year	Rate ($/kWh)	Annual Fixed Capacity Charge
1		
2		
3		
4		
5		
6		
7		
8		
9		
10		
11		
12		
13		
14		
15		
Thereafter		

Exhibit D

Termination Buyout and Termination Value

ADB is not permitted to exercise the buyout option in the first six (6) years of operation. The buyout /termination value shown for Year 6 refers to buyout / termination at the end of Year 6 / beginning of Year 7.

Year	Buyout Value	Termination Value
1	No buyout/termination option	No buyout/termination option
2	No buyout/termination option	No buyout/termination option
3	No buyout/termination option	No buyout/termination option
4	No buyout/termination option	No buyout/termination option
5	No buyout/termination option	No buyout/termination option
6	No buyout/termination option	Buyout Value + Removal Fee *
7		Buyout Value + Removal Fee *
8		Buyout Value + Removal Fee *
9		Buyout Value + Removal Fee *
10		Buyout Value + Removal Fee *
11		Buyout Value + Removal Fee *
12		Buyout Value + Removal Fee *
13		Buyout Value + Removal Fee *
14		Buyout Value + Removal Fee *
15		Buyout Value + Removal Fee *
16	0	0

Notes:

1. Buyout value: lump sum payment from ADB to the Power Provider prior to the end of the Term, to take ownership of the Generating Facility and terminate this Agreement.

2. Termination value: lump sum payment from ADB to the Power Provider upon termination of this Agreement.

3. Termination value = Buyout Value + Removal Fee

4. Removal Fee = All costs associated with the removal of the Generating Facility, if ADB elects to have the Power Provider to remove the Generating Facility upon early termination of this Agreement.

Exhibit E

Example of Hypothetical Shortfall Payment Calculation

In Year 3, the Power Purchaser consumes 7,000,000 kWh of electricity and pays the Power Provider $1,050,000 for its total annual energy use under the TOU-8 rate. Therefore the blended average annual TOU-8 rate is equal to $0.15 per kWh ($1,050,000 / 7,000,000 kWh = $0.15 per kWh). The cumulative output guarantee in Year 3 is 4,193,486 kWh. The actual delivered cumulative output is 4,100,000 kWh. The shortfall is therefore 93,486 kWh. The PPA rate is $0.14333 per kWh in Year 3. The shortfall payment paid by the Power Provider to the Power Purchaser is $624 (93,486 kWh x [$0.15 – $0.1433] = $624).

The following chart specifies the Expected Annual Production from the Generating Facility as well as the Annual Production Level at 90% and the cumulative output guarantee. The Actual Output of the Generating Facility will be measured cumulatively and compared to the cumulative output guarantee level on the third, sixth ninth, twelfth, fifteenth, and twentieth year during the cumulative output guarantee term.

System Size (kWp)			1,190
Module Degradation/Year			0.75%
Year	Estimated Solar Electricity Produced (kWh)	Guaranteed kWh (90% of estimated output)	Cumulative Output Guarantee (kWh)
1	1,564,850	1,408,365	
2	1,553,114	1,397,802	
3	1,541,465	1,387,319	4,193,486
4	1,529,904	1,376,914	
5	1,518,430	1,366,587	
6	1,507,042	1,356,338	8,293,324
7	1,495,739	1,346,165	
8	1,484,521	1,336,069	
9	1,473,387	1,326,048	12,301,607
10	1,462,337	1,316,103	
11	1,451,369	1,306,232	
12	1,440,484	1,296,435	16,220,377
13	1,429,680	1,286,712	
14	1,418,958	1,277,062	
15	1,408,315	1,267,484	20,051,635

www.ingramcontent.com/pod-product-compliance
Lightning Source LLC
Chambersburg PA
CBHW041120280326
41928CB00061B/3465